AF863134

Thermodynamics of Soil Nutrient Bioavailability

Kodoth Prabhakaran Nair

Thermodynamics of Soil Nutrient Bioavailability

Sustainable Soil Nutrient Management

Springer

Kodoth Prabhakaran Nair
Villament G3
C/o Dr. Mavila Pankajakshy
Malaparamba, Kozhikode, Kerala, India

ISBN 978-3-030-76816-4 ISBN 978-3-030-76817-1 (eBook)
https://doi.org/10.1007/978-3-030-76817-1

© The Editor(s) (if applicable) and The Author(s), under exclusive license to Springer Nature Switzerland AG 2021
This work is subject to copyright. All rights are solely and exclusively licensed by the Publisher, whether the whole or part of the material is concerned, specifically the rights of translation, reprinting, reuse of illustrations, recitation, broadcasting, reproduction on microfilms or in any other physical way, and transmission or information storage and retrieval, electronic adaptation, computer software, or by similar or dissimilar methodology now known or hereafter developed.
The use of general descriptive names, registered names, trademarks, service marks, etc. in this publication does not imply, even in the absence of a specific statement, that such names are exempt from the relevant protective laws and regulations and therefore free for general use.
The publisher, the authors, and the editors are safe to assume that the advice and information in this book are believed to be true and accurate at the date of publication. Neither the publisher nor the authors or the editors give a warranty, expressed or implied, with respect to the material contained herein or for any errors or omissions that may have been made. The publisher remains neutral with regard to jurisdictional claims in published maps and institutional affiliations.

This Springer imprint is published by the registered company Springer Nature Switzerland AG
The registered company address is: Gewerbestrasse 11, 6330 Cham, Switzerland

I dedicate this book, written and published, under very trying times, due to the global covid pandemic, to Pankajam, my wife, who is all to me, and also to the memory of my late parents, my father, Kuniyeri Pookkalam Kannan Nair, a very illustrious Police Officer, who served the British Police and who was decorated with the King George V medal for bravery and honesty, and my mother, Kodoth Padinhare veetil

Narayanai Amma, daughter of the aristocratic Kodoth family of North Malabar, Kerala State, India, who I lost at a very young age, but whose endearing love and blessings made me what I am today

A Word of Appreciation

This book has been compiled at the most difficult phase of my professional life, with so many constraints coming in the way, paramount among which is the covid pandemic. I am very pleased that finally the task has been completed and I wish to add this:

At the outset, I place on record my heartfelt gratitude to Ms. Margaret Deignan, Senior Editor and Publisher, Springer, who, from the very start of the idea for this book, has been a constant source of inspiration, encouragement and guidance to me. I owe her a deep debt of gratitude.

Mr. Gideon Philip, Project Coordinator (Books) has been most conscientious in advising me whenever a bottleneck arose in production, and, his colleague Ms. T. Suganthi, Project Manager and her team did a remarkable job managing all the crises. I am very sincerely thankful to both.

Contents

Chapter 1
Introduction

Abstract The chapter introduces the reader to the "concept of soil". The discussion would focus on the centrality of soil, as the source of sustenance to all life—humankind, animals, and plants. "The fabric of human life is woven on earthen looms—it everywhere smells of the clay", as quoted by Roy W. Simoson, a distinguished soil scientist, sums up the centrality of soil. The author of this book would consider soil as the "Soul of Infinite Life", substituting the first letter of each word—Soul, Of, Infinite, Life- in the phrase. In short, it is soil that all life comes forth, and, it is soil that all life goes back to, finally.

India's great President late Dr. A.P.J. Abdul Kalam, launching the book "ISSUES IN NATIONAL AND INTERNATIONAL AGRICULTURE", authored by Professor Kodoth Prabhakaran Nair, in Raj Bhavan, Chennai, Tamil Nadu, India

© The Author(s), under exclusive license to Springer Nature Switzerland AG 2021
K. P. Nair, *Thermodynamics of Soil Nutrient Bioavailability*,
https://doi.org/10.1007/978-3-030-76817-1_1

Many years ago, in one of the early editions of Advances in Agronomy, Roy W. Simonson writing a chapter entitled "Concept of Soil," noted: "Someone has said that the fabric of human life is woven on earthen looms—it everywhere smells of the clay." More than three decades later, we agronomists and soil scientists have come very far in our understanding of "the fabric of human life" which "everywhere smells of the clay." That "the fabric of human life" which is so very intimately linked to soil has dramatically changed is beyond dispute. Yet, there is no denying the fact that this "fabric of human life" will always be linked to the soil which is "the pragmatic, the reality, the entity that dictates much of what societies can do" (Boul 1994). Soil, in my opinion, is that invaluable gift of God to life on planet Earth and can aptly be termed "The Soul of Infinite Life." Though the basic concept of soil, since its early description as a "thin mantle over the land surface" has vastly changed over the years, this thin mantle has always been the focal point since it is the medium for plant growth. For early man it was nothing more than a physical support for his predation. Quite likely, some areas were known to provide better footing than others, and some were to be avoided if possible. It is amazing that even after decades of research in soil science, which has provided such invaluable information on this "thin mantle over the land surface" so crucial to the existence of life, human, plant, and animal, on planet earth, this basic instinct of predation has remained unchanged. How else can we explain the disdain and callousness so often witnessed in modern societies, propelled by an insatiable greed to acquire unlimited wealth, which leads to the abuse of soil, this invaluable gift of God to man? Undoubtedly, the earliest shift in attitude toward soils must have originated at a time when man began to grow food, rather than gather it as his ancestors did. In many ways, this shift in attitude was the precursor to modern-day soil science. Though this shift must have occurred in pre-Christian times, about 9000 years ago, and focused on the inevitability of a proportionally smaller land surface supporting a larger human population, it is only in recent times that we have witnessed the magnitude of the impact of this shift in attitude on human existence. Much land has become degraded and unsuitable for agriculture since a century ago. The 1992–1993 World Resources report (Stammer 1992) from the United Nations on the status of the world soils contains very alarming conclusions. For example, nearly 10 million ha of the best farm lands of the world have been so ruined by human activity since World War II that it is impossible to reclaim them. Over 1.2 billion ha have been seriously damaged and can be restored only at a great cost. This loss in soil capability could mean that there will be enormous food shortages in the next 20–30 years and, as is but natural, the people of disadvantaged nations will suffer the most. Many factors have contributed to this alarming state of affairs, one of the prime factors being "high-input agriculture," or more specifically chemical agriculture, euphemistically known as the "green revolution," where unbridled use of chemical fertilizers led to soil ruination. Punjab, the "cradle of green revolution" in India is a testimony to this sad state of affairs, where unbridled use of chemical fertilizers to boost the yield of dwarf wheat and rice varieties has led to soil degradation, loaded the ground water with high amounts of fertilizer residues (especially nitrates from urea) that it is no longer potable, led to soil salinity, dried aquifers, and vanishing biodiversity due to continuous monoculture of wheat and rice. There are hundreds of acres of land stretches that have become barren, where once stood lush wheat and rice fields. Crop yields have plateaued or drastically declined. Two-thirds of the seriously eroded land is in Asia and Africa. About 25% of the cropped land in Central America is moderately to severely damaged. In North America, this is a small percentage—only 4.4%. Since the time of this "green revolution," food production has declined dramatically in 80 developing countries in the past decade. Soil degradation is the major factor. Nearly 40% of the world's farming is done on very small parcels of 1 ha or less (Robison et al. 1981). Ignorance and poverty characterize this situation. Yet, emphasis on agriculture has been confined mostly to large-scale farming. Large-scale farming, grand projects at huge costs and huge profits, have been the order of the day for many decades. In a lighter vein, it can be said that even the "lebensraum" concept of Adolf Hitler had an echo in the inevitability of this modern-day fact. What else can justify the ruthless conquest of vast territories of land by this master strategist who set out to conquer the world—or, more appropriately, the world's soils?

Despite the complexity of soil science and the emergent soil management practices, the basic concept of soil as a medium of plant growth can be expected to persist for an indefinite length of time. But it is becoming increasingly clear that the earlier views on soil as merely the "supportive medium" for plant growth is giving place to newer ones on "managerial concepts" of this supportive medium. This is amply illustrated by the shift in focus from the green revolution phase of the 1960s to mid-1970s where application of increasing quantities of soil inputs such as fertilizers and pesticides was emphasized, to the "sustainable agriculture" phase from the early 1980s to the present (probably to continue?). Sustainable agriculture places more reliance on biological processes by adopting genotypes to adverse soil conditions, enhancing soil biological activity and optimizing nutrient cycling to minimize external inputs, such as fertilizers, and maximize their efficiency of use. In fact, the paradigm of the earlier phase has given way to the emergent new paradigm (Sanchez 1994) and this is clearly reflected in the dialogue of the world leaders during the Earth Summit in 1991 in Rio de Janeiro, Brazil, where Agenda 21 has incorporated six chapters on soil management issues (Keating 1993). The focus of this review will be on the second paradigm inasmuch as prescriptive soil management is concerned with regard to understanding soil nutrient availability and its efficient management in crop production.

1.1 Laws Governing Sustainable Soil Management

The challenge of doubling world's food grain production by 2030 is a lot more daunting than imagined, because, the per capita arable land and renewable fresh water resources have decreased dramatically over the years. The highly soil extractive and environmentally very damaging farming methods, practiced over the last more than half a century, euphemistically called the "green revolution", has left most of the soil resources of Southern Asia, in particular, of a country like India, totally degraded. For instance, of the total geographical area of India of 328.73 million hectares, as much as 120.40 million hectares are degraded, that is almost a third of the total geographical area, thanks to the green revolution. North Indian states like Punjab, known as the "cradle of the green revolution" has thousands of acres where not even a blade of grass will grow, unless, huge investments are made in soil ameliorative techniques. Inherent soil fertility has vastly deteriorated. The soils of *Kuttanad,* in the State of Kerala, known as the "Rice Bowl" of Kerala are so degraded, thanks to the unbridled use of chemical fertilizers to prop up the green revolution, that soil pH has come down from a healthy value of 6–6.5 in the pre-green revolution days to 2–3 now; soils have turned very acidic and there is yield plateauing or drastic yield decline. The global warming (Nair 2019b) is now an added worry. Close to 35% of global warming is due to the unbridled use of nitrogenous fertilizers, in particular, urea, done to prop up the green revolution, which while hydrolyzing in soil releases the nitrous oxide (N2O) which captures the radiated heat in the stratosphere leading to global warming (Nair 2019a, b).

1.2 Basic Principles of Sustainable Soil Management

According to Rattan Lal (2009) there are ten basic laws that govern sustainable soil management, and, they are the following:

Law #1:

Soil resources are unequally distributed among biomes and geographic regions. Highly productive soils in favorable climates are finite and often located in regions of high population density and have already been converted to managed ecosystems; managed crop land, grazing land and pasture, forest and energy plantations.

Law#2:

Most soils are prone to degradation by Land misuse and soil management. The highly soil extractive "green revolution", in Punjab State, India, illustrates this vividly (Nair 2019a, b). Anthropogenic factors leading to soil degradation are driven by desperate situations and helplessness in the case of resource poor farmers and smaller landholders, and, greed, short-sightedness, poor planning and cutting corners for quick profits in the case of large-scale and commercial farming enterprises (Nair 2019a, b).

Law #3:

Accelerated soil erosion and decline in soil quality by other degradation processes depend more on "how" rather than "what" crops are grown. Productive potential of farming systems can only be realized when implemented in conjunction with restorative and recommended soil and water management practices. Sustainable use of soil depends on the judicious management of both on-site and off-site inputs. Indiscriminate and excessive use of tillage, irrigation and fertilizers can lead to as much as or even more degradation than none or minimal use of these technologies (Nair 2019a, b).

Law #4:

The rate and susceptibility of soil to degradation increase with increase in mean annual temperature and decrease in mean annual precipitation. All other factors remaining the same, soils in hot and arid climates are more prone to degradation than and desertification those in cool and humid ecoregions. However, mismanagement can lead to desertification even in arctic climates, for example, Iceland.

Law# 5:

Soil can be a source or sink of green house gases, for example, CO2, CH4, and N2O depending on land use and management (Nair 2019a, b). Soil is a sink of atmospheric CO2 under those land use and management systems which create a positive C budget and gains exceed the losses. Soil is a source of atmospheric CO2 when the ecosystem C budget is negative and losses exceed gains. Soils are a source of radiatively-active gases with extractive farming (like the green revolution, Nair, 2019a, b) which create a negative nutrient budget and degrade soil quality (Nair 2019a, b), and a sink with restorative land use and judicious management practices which create positive C and nutrient budgets (Nair and Mengel 1984) and conserve soil and water while improving soil structure and tilth.

Law#6:

Soils are non-renewable over a human time frame of decadal or generational scales, but, are renewable on a geological scale (centennial/millennial). With the increase in human population, projected to be 10 billion by 2100, restoring degraded and desertified soils over a centennial-millennial scale is not an option. Because of heavy demands on finite resources, soils are essentially a non-renewable resource.

Law#7:

Soil's resilience to natural and anthropogenic perturbations depends on it physical, chemical and biological processes. Favorable chemical and biological processes enhance resilience only under optimal soil physical properties, for example, soil structure, and tilth, processes, for example, aeration, water retention and transmission, and edaphological environments, for example, soil temperature.

Law 8 #:

The rate of restoration of the soil organic matter pool is extremely slow. While that of depletion is often very rapid. In general, restoration occurs on a centennial time scale and depletion on a decadal time scale. The rate of restoration and degradation processes may differ in order of magnitude.

Law #9:

Soil structure, similar to an architectural design of a functional building, depends on stability and continuity of macro, meso-and micropores which are the sites of physical, chemical, and biological processes that support soil's life support functions. Sustainable management systems, site-specific as these are, enhance stability and continuity of pores and voids over time and under diverse land uses.

Law #10:

Sustainable management of agricultural ecosystems implies an increasing trend in net primary productivity per unit input of off-farm resources along with improvement in soil quality and ancillary ecosystem service, such as, increase in the ecosystem C pool, improvement in quality, and quantity, of renewable fresh water resources and increase in biodiversity.

Soil resources can never be taken for granted. Extinct are the once thriving civilizations, for example, Mayan, Incas, Indus, Mesopotamia, which chose to ignore their soil resources. Given the importance to human survival and dependence on terrestrial life, soil quality must be improved, and, restored. Soils must be transferred to the next generation in a better state than when received from the previous one.

References

Boul SW (1994) Soil, society's yoke to the earth. In: Proc. XV of the ISSS symposium, Acapulco, Mexico, July 10–16, vol 1, pp 89–104

Keating M (1993) The Earth Summit's agenda for change: a plain language version of Agenda 21 and the other Rio Agreements. Centre for Our Common Future, Geneva

Lal R (2009) Laws of sustainable soil management. In: Lichtfouse E, Navarrete M, Debaeke P, Souchere V, Alberola C (eds) Sustainable Agriculture. Springer, Dordrecht, pp 9–12

Nair KPP (2019a) Intelligent soil management for sustainable agriculture. The nutrient buffer power concept. Springer, p 389

Nair KPP (2019b) Combating global warming. The role of crop wild relatives for food security. Springer, p 120

Nair KPP, Mengel K (1984) Importance of phosphate buffer power for phosphate uptake by rye. Soil Sci. Soc. Am. J. 48:92–95

Robison LR, Johnston NP, Hill JM (1981) Small-scale agriculture, an untapped giant. Benson Inst 4(2):4–7

Sanchez PA (1994) Tropical soil fertility research: towards the second paradigm. In: Proc. XV ISSS symposium, Acapulco, Mexico, July 10–16, vol 1, pp 89–104

Stammer LB (1992) Study finds serious harm to 10 percent of world's best soil. Los Angeles Times (quotations from World Resources Institute of U.N. Environmental Program)

Chapter 2
Efficient Plant Nutrient Management—The Key Factor in Sustainable Soil Management

Abstract Agricultural systems differ from natural systems in one fundamental aspect; while there is a net outflow of nutrients by crop harvests from soils in the first, there is no such thing in the second. This is because nutrient losses due to physical effects of soil and water erosion are continually replenished by weathering of primary minerals or atmospheric deposition. Hence, the crucial element of sustainability of crop production is the nutrient factor. The chapter discusses details with this objective in focus.

Agricultural systems differ from natural systems in one fundamental aspect: while there is a net outflow of nutrients by crop harvests from soils in the first, there is no such thing in the second (Sanchez 1994). This is because nutrient losses due to physical effects of soil and water erosion are continually replenished by weathering of primary minerals or atmospheric deposition. Hence, the crucial element of sustainability of crop production is the nutrient factor. But, of all the factors, the nutrient factor is the least resilient (Fresco and Kroonenberg 1992). The thrust of high-input technology, the hallmark of the "green revolution," in retrospect, or the moderation by low input technology, the foundation stone of sustainable agriculture, in prospect, both dwell on this least-resilient nutrient factor. If the pool of nutrients in the soil, both native and added, could be considered as the "capital," efficient nutrient management might be analogous to raising the "interest" accrued from this capital in such a way that there is no great danger of the erosion of this capital. Hence, sound prescriptive soil management should aim at understanding the actual link between the "capital" and the "interest" so that meaningful management practices can be prescribed.

2.1 Soil Tests and Nutrient "Availability"

It is universal knowledge that soil tests are the basis for predicting nutrient "availability." There are, perhaps, as many soil tests for each nutrient as there are nutrients. This review will not dwell on the merits or demerits of any single soil test or

© The Author(s), under exclusive license to Springer Nature Switzerland AG 2021
K. P. Nair, *Thermodynamics of Soil Nutrient Bioavailability*, https://doi.org/10.1007/978-3-030-76817-1_2

group of them. Suffice it to say that fertilizer recommendations traditionally are made at the point where marginal revenues equal marginal costs, which involve some positive synergism (DeWit 1992). The most common result of this approach is the vast build-up in the soil nutrient pool in intensively cultivated soils (Whitmore and van Noordwijk 1994). Data in Table 2.1 indicate positive balances (in kg/ha/year) for N(61), P(23), and K(37) in intensive crop production systems (Frissel 1978).

Over several decades, such positive balances can lead to a huge build-up of the nutrient capital, especially in the case of high-input, intensive agricultural systems as in the case of many European, North American, and Scandinavian countries. A dangerous consequence of such huge soil build-up is nutrient contamination of groundwater to such extremes that "environmental soil tests" become necessary to assess critical limits of nutrient pollution (Sharpley et al. 1993). Nitrogen is a prime candidate for this scenario, especially in the temperate zone. At the other end of the spectrum are the marginal areas of the tropical zone where inadequate replenishment of nutrient removal by crop growth and also nutrient loss by soil and water erosion has left that capital "in the red." Initially fertile Alfisols of much of Africa with subsequent severe depletion of N and P (Yates and Kiss 1992) are an example of this nutrient "bankruptcy." Either way, contemporary soil tests are the basis on which prescriptive management practices are formulated.

2.2 Rating Soil Tests to Define Nutrient Availability and a Fertility Index

Most soil test laboratories around the world use some kind of "rating system" to evaluate soil test values. These rating systems invariably use qualitative terms such as "low," "medium," or "high" to describe the availability of a specific plant nutrient. Admittedly, these terms denote different meanings in the context of availability of a particular plant nutrient and, at best, are empirical terminologies. This problem

Table 2.1 Nutrient balance (kg/ha/year) in intensively managed arable soils

	N	P	K
Inputs			
Fertilizers	156	39	119
Other	32	–	9
Total	188	39	128
Outputs			
Harvest	103	16	91
Removal	24	–	–
Other			
Total	127	16	91
	Balance: 61	23	37

After Frissel (1978)

has been recognized by researchers over the years. Morgan (1935) suggested a scale of 1–10 with 8 equal to the point of no response. Bray and Kurtz (1945) used relative yield or percentage sufficiency to describe the degree of deficiency, with 100 defined as the point of no response. The index below 100 follows the curvilinear relationship between soil test values and yield without the addition of the element. Above 100, the index displays a straight-line relationship indicating the relative margin of adequacy or the proximity to an excessive level. To eliminate the need for a percent sign, the values are referred to as "Fertility Indices" and they are reported to the nearest multiple of 10 from 0 to 9990 (Cope and Evans 1985). In addition to ratings, most laboratories use some method of reporting results more precisely, mainly for use by farmers in record keeping and monitoring soil fertility. Some report kg/ha, lb./a, or ppm extracted, but these would be confusing to farmers, because each element has a different level for a specific degree of adequacy (Cope and Evans 1985). For instance, the adequate or critical level for one soil may be 25 ppm P, 120 ppm K, 200 ppm Ca, and 30 ppm Mg. Adequate levels in other soils and from other extracting procedures would be different for each element (Cope and Evans 1985).

Despite the fact that a number of soil tests and others such as Diagnosis and Recommendation Integrated Systems are in vogue to predict nutrient availability, it must be said that a universal picture is yet to emerge in this field with regard to precise availability prediction. This is primarily because a soil test in the laboratory can never simulate plant root absorption of a nutrient in a field soil, though most of the time the assumption is that it does. In the final analysis, it is the plant and plant alone which will decide whether or not the nutrient is available. This review examines the question of whether a better and more reliable alternative exists.

References

Bray RH, Kurtz LT (1945) Determination of total organic and available form of phosphorus in soils. Soil Sci 59:39–45

Cope JT, Evans CE (1985) Soil Test Adv Soil Sci 1:201–228

DeWit CT (1992) Resource use efficiency in agriculture. Agric Syst 40:125–151

Fresco LO, Kroonenberg SB (1992) Time and spatial scales in ecological sustainability. Land Use Policy, July 1992, pp 155–167

Frissel MJ (ed) (1978) Cycling of mineral nutrients in agricultural ecosystems. Elsevier, Amsterdam

Morgan MF (1935) A universal soil testing system. Conn Agric Exp Stn Bull 372

Sanchez PA (1994) Tropical soil fertility research: towards the second paradigm. In: Proc. XV ISSS symposium, Acapulco, Mexico, July 10–16, vol 1, pp 89–104

Sharpley ANI, Pierzynki G, Sims JT (1993) Innovative soil phosphorus availability indices: assessing inorganic phosphorus. In: Agronomy abstracts. American Society of Agronomy, Madison, WI

Whitmore AP, van Noordwijk M (1994) Bridging the gaps between environmentally acceptable and agronomically desirable nutrient supply. In: Glen D (ed) Proceedings of the long Ashton symposium: agriculture in the 21st century

Yates RA, Kiss A (1992) Using and sustaining Africa's soils. In: Agricultural and rural development series no. 6. World Bank, Washington, DC

Chapter 3
The Thermodynamics of Soil Nutrient Bio Availability

Abstract In any soil nutrient management approach that is sound ad reproducible, one must start with the basic understanding of the chemical environment of plant roots. When one considers this, the first term one comes across is the "soil solution", because, the plant root is bathed in it. The Soil Science Society of America (Soil Sci Soc Am Proc 29:330–351, 1965) defines soil solution as "the aqueous liquid phase of the soil and its solutes consisting of ions dissociated from the surfaces of the soil particles and other soluble materials". The chapter would develop the central argument in the book on the basis of the thermodynamics involved conforming to mechanical-mathematical models, to derive the "Buffer Power Concept".

3.1 Basic Concepts

In any nutrient management approach that is sound and reproducible, one must start with a basic understanding of the chemical environment of plant roots. When we consider this, the first term that we come across is the "soil solution", because the plant root is bathed in it and most affected by its chemical properties. The Soil Science Society of America (1965) defines soil solution as "the aqueous liquid phase of the soil and its solutes consisting of ions dissociated from the surfaces of the soil particles and other soluble materials". Adams (1974) has given a simple definition: "The soil solution is the aqueous component of a soil at field moisture contents". Perhaps it is important to emphasized here that much of contemporary soil testing has considered a soil extract as synonymous with the soil solution. Since soil extraction is supposed to simulate plant root extraction, it is pertinent to consider the chemical environment of the root, though briefly, from this angle. It is worth noting that the chemical environment of roots in natural soil systems is so obviously complex that both soil scientists and plant physiologists have been unable to provide a precise definition. F this complex chemical system is to be accurately quantified, thermodynamic principles will need to be used to evaluate experimental data. Even then, the limitations are obvious, as in the case of K where the thermodynamic investigations are quite often inapplicable under field conditions. This is because, although a quasi-equilibrium in K exchange can be achieved in the laboratory, these conditions are seldom, if ever, attained under field conditions (Sparks

© The Author(s), under exclusive license to Springer Nature Switzerland AG 2021

K. P. Nair, *Thermodynamics of Soil Nutrient Bioavailability*,
https://doi.org/10.1007/978-3-030-76817-1_3

1987). Agricultural soils are, for the most part, in a state of disequilibrium owing to bot fertilizer input and nutrient uptake by plant root. It thus appears that a universal and accurate definition of a root's chemical environment awaits the proper application of thermodynamics for the root's ambient solution (Adams 1974) or even kinetics.as in the case of K (Sparks 1987), where thermodynamics has been found inadequate.

Soil extractions with different extractants provide a second approach in defining the root's chemical environment. This approach has been particularly successful in understanding cases like P solubility, soil acidity, and K fixation. However, this approach also fails to define precisely the root's chemical environment. Hough this approach also suffers from deficiencies, such as the extractants removing arbitrary and undetermined amounts of solid-phase electrolytes and ions (or the extractants causing precipitation of salts or ions from the soil solution) and the soil-plant interrelationship defined in terms of the solid phase component of the soil, even though the solid phase is essentially inert except as it maintains thermodynamic equilibria with the solution phase (Adamas 1974), the latter part could be researched more to understand how the solid phase-solution phase equilibria can be interpreted to give a newer meaning to quantifying nutrient bio availability. It is in this context that the role of the plant nutrient's "buffer power" assumes crucial importance.

The close, almost linear, relationship in a low concentration range of <0.5 mM for $NO_3 - N$, $NH_4^+ - N$, K^+, $H_2PO_4^-$, and HPO_4^{2-}, which has been established by numerous solution culture experiments, can be quantitatively defined by the equation

$$U = 2\pi r \alpha C_r \tag{3.1}$$

where U is the uptake of a 1-m root segment, r is the root radius, C_r is the concentration of the ion at the root surface, and α is the root absorbing power (Mengel 1985). The metabolic rate of the root determines its absorbing power. A high root absorbing power would imply that a relatively high proportion of nutrient ions coming in contact with the root surface is absorbed and vice versa. The nutrient ion concentration at the root surface (C_r) depends on α since a high root absorbing power tends to decrease Cr; it also depends on the rate of movement from bulk soil toward the root (Mengel 1985). Diffusion and/or mass flow controls this movement. But it is now established that nearly 95% of this movement for nutrients such as P, K and Zn (among heavy metals) and, possibly NH_4^+, is by way of diffusion. When root uptake of an ion species is less than its movement toward it, accumulation of the ion species on the root surface is bound to occur, as has been shown to be the case of Ca^{2+} where mass flow contributes to this accumulation (Barber 1974). The diffusive path for ions, such as P and K, which plant roots take up at high rates, but, which are in low concentration in the soil solution near the root, is the concentration gradient. In a sense, the effective diffusion coefficient which quantifies the diffusive path and the buffer power are analogous because the diffusive flux across the root surface is integrally related to the nutrient buffer power. This has been shown to be true in the case of P, where a highly significant positive correlation between the two was found to exist in 33 soil samples obtained from experimental sites in the United States and

Canada (Kovar and Barber 1988). However, in a routine laboratory setup, it is far easier to measure the buffer power of the plant nutrient in question than the effective diffusion coefficient and this review will further focus on the question of how buffer power can be quantified without recourse to cumbersome analytical techniques and how its integration into routine soil test data will considerably improve predictability of nutrient uptake.

3.2 Measuring the Effective Diffusion Coefficient of a Nutrient (Nutrient Buffer Power) and Its Importance in Affecting Nutrient Concentrations on Root Surfaces

The ability to predict the mobility of dissolved chemicals, such as fertilizers, in the soil is of considerable interest and value in managing fertilizer applications. Soil testing, in its essence, aims to achieve this. While modeling transport and retention of ions from thermodynamic (Selim 1992), kinetic (Sparks 1989), and mechanistic (1984) angles could be very informative, the importance of translating this information into practically feasible procedures in crop production calls for an understanding not only of the basic concepts, but, their intelligent application as well. In a dynamic state of plant growth, the concentration of any nutrient on the root surface is nearly impossible to precisely measure since both the nutrient in the plant tissue and the root absorbing power, which directly affects it, change quickly due to root metabolic processes. The inability of even mechanical-mathematical models to accurately predict nutrient influx rates has recently been highlighted (Lu and Miller 1994). Hence, if an effective soil testing procedure is to be devised for a nutrient, which is an alternative to defining the plant root's chemical environment, one must resolve the problem of quantifying the nutrient concentration on the root surface indirectly, even if impossible to resolve it directly, for the reasons mentioned above.

Using Fick's first law,

$$F = -D(dC/dx)$$

Where F is the flux, dC/dx is the concentration gradient across a particular section, and D is the diffusion coefficient, Nye (1979) has suggested that the formula can be applied to both ions and molecules. The negative sign for D implies net movement from high to low concentration. Although for molecules in simple systems like dilute solutions D may be nearly constant over a range of concentrations, for ions in complex systems like soils and clays D will usually depend on the concentration f the ion, and on that of other ions as well (Nye 1979). Nye (1979) has further suggested that though Fick's first law may be derived from thermodynamic principles in ideal systems, in a complex medium such as the soil, the above equation may be regarded as giving and operational definition of the diffusion coefficient. Thus, Nye (1979) defines diffusion coefficient as

$$D = D_l \theta f_l \left(dC_l / dC \right) + D_E \tag{3.2}$$

Where D_l is the diffusion coefficient of the solute in free solution, θ is the fraction of the soil volume occupied by solution and gives the cross section for diffusion, f_l is an impedance factor, C_l is the concentration of the solute in the soil solution, D_E is an excess term which is zero when the ions or molecules on the solid have no surface mobility, but, represents their extra contribution to the diffusion coefficient when they are mobile. D_E can generally be neglected since only in rare instances will it play any role in diffusion of plant nutrient ions in soil (Mengel 1985). From the point of view of nutrient bio availability, dC_l/dC, which represents the concentration gradient, assumes crucial importance, as we shall see below.

The term dC_l/dC, were C_l is the concentration of the nutrient ion in the soil solution and C is the concentration of the same ion species in the entire soil mass, assumes considerable significance in lending a practical meaning to nutrient bio availability. If we ascribe the term "capacity" or "quantity" to C and "intensity" to C_l, we have in this term an integral relationship between two parameters that may crucially affect nutrient bio availability. Since the concentration gradient of the depletion profile of the nutrient in the zone of nutrient uptake depends on the concentration of the ion species in the entire soil mass (represented by "quantity" or "capacity") in relation to the rate at which this is lowered on the plant root surface by uptake (represented by "intensity"), it could be argued that a quantitative relationship between the two should represent the rate at which nutrient depletion and/or replenishment in the rooting zone should occur (Nair 1984a, b). This relationship has been functionally quantified by Nair and Mengel (1984) for P in eight widely differing Central European soils (Table 3.1, see below), and, the term dC_l/dC has been referred to as the "nutrient buffer power".

Nair and Mengel (1984) used electro-ultra-filtration (EUF) to quantify C_l while using an incubation and extraction technique to quantify C. For P, C was found to closely approximate isotopically exchangeable P (Keerthisinghe and Mengel 1979), but in the experiments conducted by Nair and Mengel (1984), it was estimated by the extraction of incubated soil with an extractant which was a mixture consisting of 0.01 M Calcium lactate +0. 1 Calcium acetate +0.3 M acetic acid at pH 4.1. The extractant exchanges adsorbed phosphate and dissolves Ca phosphates except apatites; the method known as the "CAL method", developed by Schuller (1969), is now widely used in central Europe. In the case of K^+ and NH_4^+ – N C denotes the concentration of exchangeable, and to some extent non exchangeable fractions (Mengel 1985). Since very low concentration in the range of 2.0 μ M may be attained on the root surface for both P and K (Claassen and Barber 1976; Claassen et al. 1981; Hendriks et al. 1981), Nair and Mengel (1984) had to use EUF technique to quantify C_l. Thus, the nutrient depletion around the roots which is caused by the diffusive flux of the nutrients toward the root surface is related to both the quantity and the intensity parameters, a quantifiable relationship between both represents the

Table 3.1 Comparison of P effective diffusion coefficient (P buffer power) of eight widely differing Central European soils determined by two different techniques

	Regression		"r"	
Soil	(1)	(2)	(1)	(2)
Benzheimer Hof	Y = 18.8x + 7.94	Y = 0.23x + 8.98	0.912	0.995
Hungen	Y = 38.2x −1.03	Y = 4.32 + 0.25x	0.967	0.997
Oldenburg B6	Y = 49.8x + 0.52	Y = 0.72 + 0.26x	0.994	0.999
Wolfersheim	Y = 70.3x + 0.03	Y = 0.11 + 0.27x	0.998	0.983
Obetrshausen	Y = 70.5x + 2.66	Y = 2.89 + 0.30x	0.966	0.998
Oldenburg B3	Y = 73.6x + 2.07	Y = 0.61 + 0.31x	0.994	0.997
Klein-Linden	Y = 75.0x + 0.38	Y = 1.81 + 0.32x	0.999	0.991
Gruningen	Y = 75.4x + 0.89	Y = 3.62 + 0.36x	1.000	0.996

Note: The *b* values in the regression functions represent the P effective diffusion coefficient (P buffer power of each soil). In regression function (1) (after Nair and Mengel, 1984) Y = CAL-P (Schuller's method) and in regression function (2) (after Nair 1992a, b) Y = the author's method. x in both refers to the elctro-ultra-filtrable P. Note the very high "r" values in all the cases. The soils are arranged in their sequential increase in effective P diffusion coefficient (P buffer power)

buffer power specific to the nutrient and the soil. A growing root will at first encounter a relatively high concentration of P, which is in the range of the concentration of the bulk soil solution (Nair and Mengel 1984). As uptake continues, a depletion will occur at the root surface. This depletion profile gets flatter with enhanced nutrient uptake (Classen et al. 1981; Hendriks et al. 1981; Lewis and Quirk 1967). But, it is the capacity of the soil to replenish this depletion which ensures a supply of nutrient ions to the plant root without greatly depressing its average concentration on the root surface. It is the nutrient's effective diffusion coefficient (or buffer power) that decides these depletion and/or replenishment rates. A soil with a high P buffer power implies that the P absorbed from the soil solution is rapidly replenished. In such a case, the P concentration at the root surface decreases only slowly and mean P concentration at the root surface remains relatively high. In soils with a low P buffer power, the reverse is true, and, P concentration at the root surface is rapidly diminished and remains relatively low. This has been proved experimentally for P (Nair 1992a, b; Nair and Mengel 1984). This phenomenon holds true for Zn^{2+}, (Nair 1984a, b), K (Nair et al. 1997), and NH_4^+ and $NH_4^+ - N$ as well (Mengel 1985).

References

Adams F (1974) Soil solution. In: Carson EW (ed) The plant root and its environment. University Press of Virginia, Charlottesville, pp 441–481

Barber SA (1974) Influence of the plant root on ion movement in soils. In: Carson EW (ed) The plant root and its environment. University Press of Virginia, Charlottesville, pp 525–564

Claassen N, Barber SA (1976) Simulation model for nutrient uptake from soil by growing plant root system. Agron J 68:961–964

Claassen N, Hendriks K, Jungk A (1981) Rubidium-Verarmung des wurzelnahen Bodens durch Maispflanzen. Z. Pflanzenerna¨hr. Bodenk 144:533–545

Hendriks L, Claassen N, Jungk A (1981) Phosphatverarmung des wurzelnahen Bodens und Phosphataufnahme von Mais und Raps. Z. Pflanzenernähr. Bodenk 144:486–499

Keerthisinghe G, Mengel K (1979) Phosphatpufferung verschiedener Bö den und ihre Ver- änderung infolge Phosphatalterung. Mitt Dtsch Bodenk Ges 29:217–230

Kovar JL, Barber SA (1988) Phosphorus supply characteristics of 33 soils as influenced by seven rates of phosphorus addition. Soil Sci Soc Am J 52:160–165

Lewis DG, Quirk JP (1967) Phosphate diffusion in soil and uptake by plants III. ^{31}P- movement and uptake by plants as indicated by ^{32}P-autoradiography. Plant Soil 26:445–453

Lu S, Miller MH (1994) Prediction of phosphorus uptake by field-grown maize with the Barber–Cushman model. Soil Sci Soc Am J 58:852–857

Mengel K (1985) Dynamics and availability of major nutrients in soils. Adv Soil Sci 2:65–131

Nair KPP (1984a) Towards a better approach to soil testing based on the buffer power concept In: Proc. 6th Int. Colloquium for the Optimization of Plant Nutrition, 2–8 September, Pierre-Martin Prevel, Montpellier, France, vol 4, pp 1221–1228

Nair KPP (1984b) Zinc buffer power as an important criterion for a dependable assessment of plant uptake. Plant Soil 81:209–215

Nair KPP (1992a) Measuring P buffer power to improve routine soil testing for phosphate. Eur J Agron 1(2):79–84

Nair KPP (1992b) Soil testing procedure to economise on fertilizer use in land management. In: The spirit of enterprise—the 1993 Rolex awards. The Rolex Foundation, Geneva, pp 323–325

Nair KPP, Mengel K (1984) Importance of phosphate buffer power for phosphate uptake by rye. Soil Sci Soc Am J 48:92–95

Nair KPP, Sadanandan AK, Hamza S, Abraham J (1997) The importance of potassium buffer power in the growth and yield of cardamom. J Plant Nutr 20(7 & 8):987–997

Nye PH (1979) Diffusion of ions and uncharged solutes in soils and clays. Adv Agron 31:225–272

Schuller H (1969) Die CAL-Methode, eine neue Methode zur Bestimmung des pflanzen verfu¨ gbaren Phosphates in Bo¨ den. Z. Pflanzenerna¨hr. Bodenkd 123:48–63

Selim HM (1992) Modeling the transport and retention of inorganics in soils. Adv Agron 47:331–384

Soil Science Society of America (Committee Report) (1965) Glossary of soil science terms. Soil Sci Soc Am Proc 29:330–351

Sparks DL (1987) Potassium dynamics in soils. Adv Soil Sci 6:2–63

Sparks DL (1989) Kinetics of soil chemical processes. Academic, San Diego

Chapter 4
Quantifying the Effective Diffusion Coefficient (or Buffer Power) of Soils and Testing Its Effect on Soil Nutrient Bio Availability

Check for updates

Abstract The chapter would discuss the experimental procedures involved in the quantification of the "Buffer Power" of soil bioavailable nutrients, taking Phosphorus as the benchmark. Phosphorus is the most problematic nutrient from the point of its behavior in the soil to its consequent bioavailability to the plant root. The experimental results, discussed in the book, would refer to eight representative Central European soils, using, established analytical procedures, and, those developed by the author of the book.

4.1 Phosphorus

When the exchangeable or desorbable nutrient ions are plotted on the y-axis and the corresponding concentration of the same ions in the equilibrated solution on the x-axis, we obtain a slope, the steepness of which represents the buffer power for that nutrient ion. The buffer power of eight Central European soils widely differing in their physicochemical properties are reported in the above table (Table 3.2).

The "quantity factor" (C) in Eq. (3.2) was determined by two methods (i) by an incubation and extraction technique (Nair and Mengel 1984) and (ii) by an adsorption-desorption-equilibrium method (Nair 1992a, b) while the "intensity factor" was measured by the EUF method (Nemeth 1979). It is highly remarkable that the correlation coefficients, in all the soils investigated, between C (represented by y) and C_l (represented by x) are very highly significant as indicated by the "r" values, which in almost all the cases were close to unity, including the Gruningen soil, in which case it was unity. This clearly established the crucial fact that the techniques employed will accurately quantify the concentration gradient for P in the soils investigated when plants are grown in those sols. If this hypothesis is true, the impact of the P buffer power will clearly be reflected in the P uptake by the test plant since the P buffer power becomes the most crucial factor in predicting the P bio availability. This can be tested by correlating P uptake with the routine soil test data without integrating the P buffer power and also by integrating the P buffer power. Nair and Mengel (1984) used the multiple linear regression function $Y = a + bx + cz$,

© The Author(s), under exclusive license to Springer Nature Switzerland AG 2021
K. P. Nair, *Thermodynamics of Soil Nutrient Bioavailability*,
https://doi.org/10.1007/978-3-030-76817-1_4

where P = uptake by the test plant, x = routine soil test data and z = the P buffer power, and, a and b are constants, to test the hypothesis. The impact of the P buffer power on P uptake *vis-à-vis* P bio availability is clearly shown by the data in Table 4.1 (below).

In both cases of routine soil test data, the predictability of uptake was remarkably increased by integrating the P buffer power. The coefficient of determination, which is a measure of the precision of these computations, increased from 15% to 79% in the case of EUF-P and from 51% to 89% in the case of CAL-P. This, in effect, means that by integrating the P buffer power data into the computations, a dramatic increase (427%) in the case of EUF-P and 75% in the case of CAL-P, was obtained in the prediction of plant P uptake by integrating the P buffer power data into the computations. Obviously, the increase has been very remarkable in the case of the EUF-P. This also shows that the quantification of either the intensity factor alone (as in the case of the EUF-P soil test data) or the quantity factor alone (as in the case of the CAL-P data) is insufficient to precisely predict P bio availability. A substantial proportion of the variance in P uptake is contributed by the P buffer power. From the above arguments it follows that soils with a low buffer power for as particular nutrient requires a high initial concentration of the ions and versa. Mengel and Busch (1982) provided experimental evidence to substantiate this in the case of K^+ uptake of rye grass growing in nine soils of widely differing physico-chemical properties. This would also point out to the important fact that the relationship between buffer power for the nutrient and the "critical" or "cut off" limit where plant yield is perceptibly reduced has an inverse relationship – a soil with a low buffer power for the nutrient resulting in a higher critical concentration in the plant tissue as compared to a soil with a relatively high buffer power having a lower critical concentration in the plant tissue for the same nutrient. This has been found in the case of K (Mengel and Busch 1982) and P (Prabhakaran Nair unpublished data). Prabhakaran Nair (unpublished data) found a very highly significant negative correlation between the plant top critical concentration for P and the P replenishment rate in the soil substrate which the P buffer power represents in summer rye (*Secale cereale* L.). In the case of dry matter yield this relationship was significant, but positive. This suggests

Table 4.1 Correlation coefficients ("r") for the relationship of P uptake versus routine P soil tests without (A) and with (B) integration of the P buffer power

	A	B
P uptake versus EUF-P	0.393*	0.8870*
P uptake versus CAL-P	0.714**	0.9415**

After Nair and Mengel (1984)
Note: Electro-ultra filtrable P (EUF-P) and Ca acetate + acetic acid extractable P (CAL-P) (Schuller 1969), respectively, used routinely to characterize P bio availability in Central European soils
*Significant at 95% confidence level
**Significant at 99% confidence level

that the P buffer power controls the P depletion and/or replenishment in the rooting zone and its accurate measurement would be the key to a precise quantification of P availability. Remarkably, the P buffer power of the soils had no direct relationship to clay or humus content, or soil pH (Nair and Mengel 1984).

4.1.1 P Buffer Power Measurement Versus Soil Test P Data for Dependability of Availability Prediction

The important question on which the choice between P buffer power and/or contemporary or routine soil tests has to be made depends on what one aims to achieve inasmuch as prescriptive soil management practices are concerned. Contemporary soil tests are based on philosophies and procedures which were developed several decades ago, with essentially no change in the general approach. In order to be accurate, a soil test must measure both solution concentration of nutrients and their rates of movement through the soil. The "buffer power approach" seems to accomplish this. Virtually no soil test provides information about the most important factors which govern phosphate availability, which are phosphate concentration in the soil solution and the phosphate buffer power. In the widely used bicarbonate method (Olsen et al. 1954) or using water extraction (van der Paauw 1971), the P quantities recovered may be 10–100 times more than that found in the soil solution (Mengel 1985). Thus, all the methods which use an acid extraction extract much more phosphate from the "labile pool" (Larsen 1967). Acid extractions, in certain instances, can also extract phosphate that under normal crop growth will be found unavailable to plant roots. The discrepancy between contemporary soil tests and crop response to P application has recently been highlighted (Nair 1992a, b). In an extended study comparing eight central European soils with extremely differing physicochemical properties, Nair (1992a, b) found that the response to a wide range of P application to summer rye grass (*S. cereale* L.) has no relevance at all to routine soil test data. Three routinely used soil tests, namely, the 0.01 M CaCl2 test (Mattingly et al. 1963), the CAL-P test (Schüller 1969), and the EUF-P test (Nemeth 1979), were considered. The fluctuations in shoot P concentration with regard to P application were not at all reflected by the soil test values. Two of the soils which were at either extremes of "availability" (as characterized by the soil tests) showed totally divergent responses to P application, while the response was found to be very consistent considering the soil's P buffer power. This proves that the accuracy of routine soil tests can be vastly enhanced by integrating the corresponding P buffer power of the soil.

Buffer curves are extremely important in determining phosphate availability. The equilibration of a soluble fertilizer incorporated into a soil is not merely a question of adsorption and/or precipitation, but also of diffusion. Even if the fertilizer source is soluble and well mixed with the soil, the phosphate distribution is not homogenous, resulting in areas of high and low phosphate concentration. Since phosphate diffusion is a slow process, an even distribution brought about by diffusion takes time. The contact period between phosphate fertilizer and the soil has an impact on P availability, which has practical importance, and this can be depicted by buffer curves as shown in Fig. 4.1 (Barekzai 1984). This figure represents on the y-axis P quantity (Ca lactate soluble) and on the x-axis P intensity (electro-ultrafilterable) and shows that the solubility of the applied P fertilizer decreased considerably during the 6 weeks of incubation as compared to the time immediately after application. In fact, these curves show the P aging process and reflect the phosphate concentration of the bulk soil solution. The steepness of the curve at a given phosphate concentration reflects the buffer power and this steepness for eight widely differing central European soils was quantified by computing a regression function between P quantity (Ca lactate soluble P) and P intensity (electro-ultrafilterable P); the b values in the regression functions were calculated to represent the P buffer power (Nair and Mengel 1984). Considerable differences can be found among soils in their P buffer power; acid tropical soils invariably have a high buffer power, meaning that at a given quantity of available P, the P concentration of the soil solution is relatively low (Pagel and van Huay 1976). With soils widely differing in their P buffer power, at an identical quantity of available P, the P concentration of the soil solution can differ by a factor of 100 (Barekzai 1984).

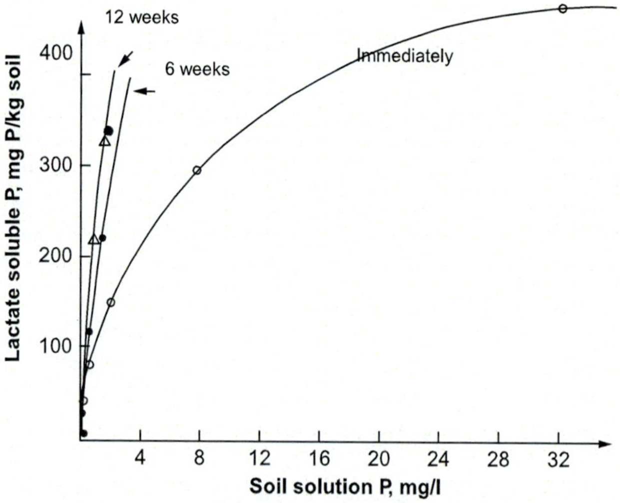

Fig. 4.1 Phosphate buffer curves of an acid soil obtained immediately after phosphate application and 6–12 weeks after phosphate application. (After Barekzai (1984))

4.1.2 P Buffer Power and Q/I Relationship

Almost half a century ago Schofield (1955), while introducing the Q/I (quantity/intensity) concept to soil P, argued that the amount of P available to plants is not necessarily dependent on the size of the labile soil P pool and suggested that the work required to remove P from the labile pool might be crucial in determining the uptake of this element by plant roots. In fact, the P buffer power might be an indirect measure of the work required to remove P from the labile pool to affect plant uptake. There has been research on both sorption and desorption Q/I relationships (Raven and Hossner 1994) and their theoretical representations are shown in Figs. 4.2 and 4.3. As soil P release is a key feature in P supply to plants, the study of P desorption Q/I characteristics might be more adequate than P sorption Q/I characteristics when evaluating plant available P and P buffer power of soils. P buffer power quantification by the sorption method would only be valid if both sorption and desorption curves overlap and if the proportion of solution P in the total desorbed P is negligible and all of the sorbed P is also desorbed (Raven and Hossner 1994). The success of the adsorption– desorption equilibrium approach in quantifying exactly the P buffer power of eight widely varying central European soils (Nair 1992a, b) lies in the above fact. However, P sorption and desorption Q/I curves generally do not coincide and sorption–desorption hysteresis is commonly observed (Barrow 1983; Fox and Kamprath 1970; Okajima et al. 1983). However, P buffer power values

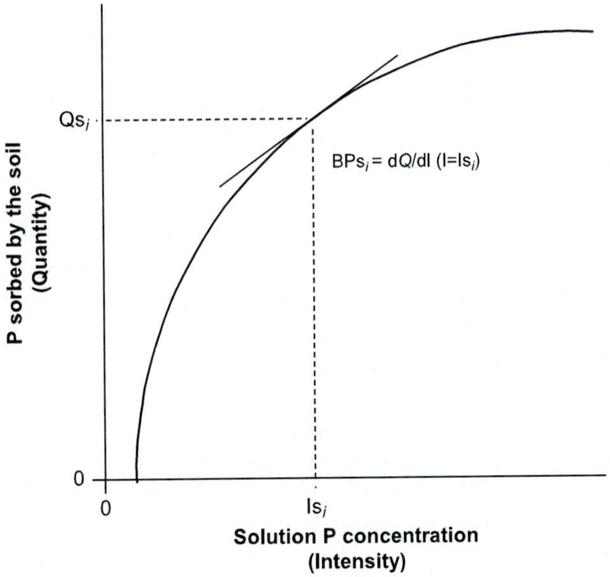

Fig. 4.2 Theoretical soil P sorption curve and related parameters (Qs_i, amount of P sorbed by the soil in situation i; Is_i, solution P concentration in situation i; and Bps_i, P buffering power at Is_i). (After Raven and Hossner (1994))

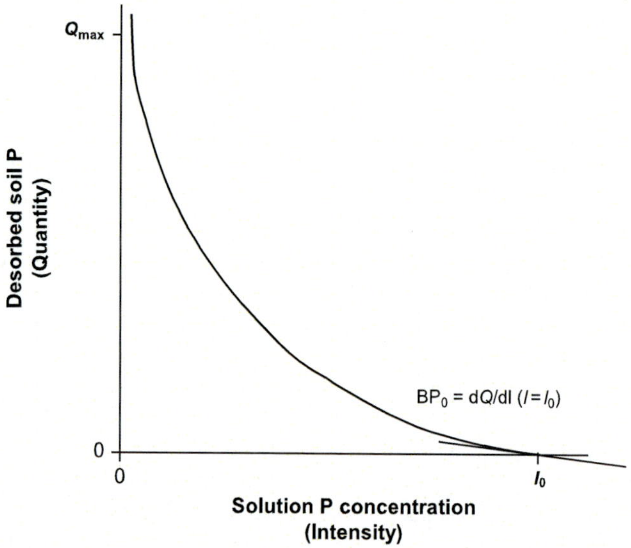

Fig. 4.3 Theoretical soil P desorption Q/I curve and related parameters (Q_{max}, maximum desorbed soil P, labile soil P, or resin-extractable P at a 4/3 soil/resin weight ratio; I_0, solution P concentration when no P is desorbed or Q ¼ 0; and BP_0, P buffering power at I_0). (After Raven and Hossner (1994))

derived from sorption curves appear to be adequate capacity indexes in terms of P uptake by plants (Bowman and Olsen 1985a; Holford 1988). The latter author, using results from 15 soils of diverse P characteristics, reported that the buffer capacities derived from P adsorption isotherms are satisfactory for characterizing P desorption, as well as adsorption, in the context of P diffusion in the soil and P uptake by plants. He further stated that during the later stages of P uptake, buffer capacities were always greater than indicated by adsorption measurements, suggesting that nonlabile P was making a significant contribution to the replenishment of the soil solution. Recent investigations (Raven and Hossner 1994) indicate that P buffer power values derived from sorption Q/I curves differ from those of desorption Q/I curves. They have reported that test plant P uptake (in this case a single corn plant in a short-term container culture) was better related to the quantity parameter derived from the desorption Q/I relationship (Qmax) as shown in Fig. 4.4. Fox and Kamprath (1970) developed a standard procedure to determine sorption curves that allow computation of the quantity of P fertilizer required to establish a desired P concentration in the soil solution. Though Beckwith (1965) tentatively suggested that adjustment to 0.2 mg P/l would be adequate for successful growth of most plants, these critical values would depend on plant species and nutritional factors (Fox 1981). Further, recourse to merely a critical solution P concentration value would neglect P buffering effect. Procedures evaluating buffering capacities of soils are often time consuming, but the question as to the need for precise documentation of soil parameters quantifying P uptake with precision for field crops, where other important factors may come into play, cannot be underestimated. Empirical

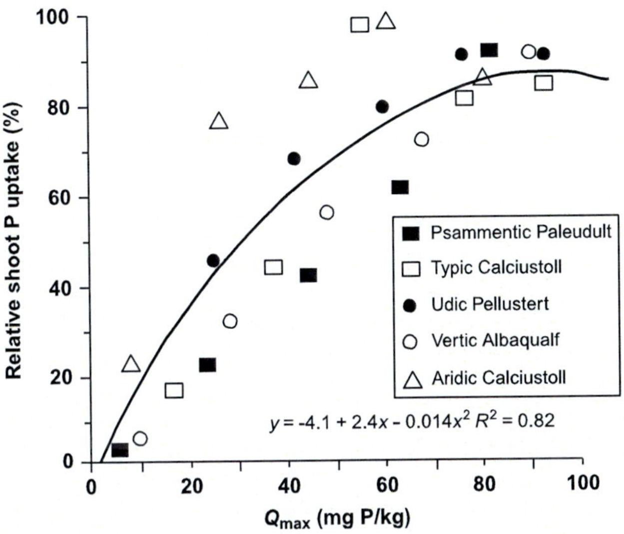

Fig. 4.4 Relationship between the relative shoot P uptake of corn plants and the maximum desorbed soil P (Q_{max}). (After Raven and Hossner (1994))

correlations based on Mitcherlich-type equations have been fairly successful in extension work, but these equations are restricted to areas with similar soil, climate, and management. Since these equations cannot accommodate soils of divergent phosphate buffering capacities, the need to include a numerical value for P buffering capacity can hardly be overlooked (Bowman and Olsen 1985b). The usefulness, for field application, of Q/I relationships obtained in the laboratory is often questioned. Q/I relationships can give an idea of the relative ease with which the "readily extractable P" will be released (buffered)—something that other conventional soil testing procedures do not give (Bowman and Olsen 1985b). Estimates can be made of the number of years the soil can supply P before deficiency levels set in. Hence, the interpretation of soil tests for soils of identical Q but with a higher P buffer power will need to be different. Another important question is whether it is critical to employ laboratory desorption isotherm procedures merely because the procedure more realistically simulates plant uptake (Bowman and Olsen 1985b). The scientific understanding of the uptake process seems more appropriately explained by the desorption isotherm process since plant uptake is basically a desorption process. However, Bowman and Olsen (1985b) suggest that where field responses are concerned, adsorption isotherms are more appropriate to standardize P buffer power of soils. The work of Nair and Mengel (1984) and Nair (1992a, b) suggests that both adsorptive and desorptive procedures are appropriate, the crucial question being how accurately these processes are determined in the laboratory and by plant culture techniques.

4.2 Case Studies from Asian Soils

4.2.1 *Potassium*

As discussed earlier with regard to P, the dynamics of K availability follows the same pattern as that of P, especially in the range of low concentrations. Beckett (1971) has used the activity ratio for K^+ and Ca^{2+} to determine K availability. Since interlayer K would play an important role in K availability, it would be more logical to consider K buffer power in determining K availability. Routinely, it is the ammonium acetate extractant that is widely used to characterize K availability. The reason that this may not be suitable to characterize exchangeable K is that in the routine extraction, only the top layer is extracted, while interlayer K from which deep-rooted plants can feed is ignored (Nair et al. 1997). There is extensive evidence to substantiate this (Mengel 1985). The importance of interlayer K in the nutrition of deep-rooted and perennial crops, such as cardamom (Elettaria cardamomum M.), the world's most valuable spice crop, has been highlighted recently by Nair et al. (1997).

4.2.2 *The Importance of K Buffer Power Determination in Predicting K Availability to Perennial Crops*

As in the case of P, the K buffer power assumes great importance in predicting K availability, especially with regard to deep-rooted and perennial crops. With annual or biennial crops, K availability has been studied with reference to the exchangeable K. However, with perennial and deep-rooted crops non-exchangeable and interlayer K play a crucial role in K availability. Three soil parameters that control the rate of K supply to plant roots which have been used for predicting K absorption by plants are the K intensity in the soil solution, the K buffer power, and the effective diffusion coefficient (Beckett 1971; Claassen et al. 1986; Mengel and Kirby 1980). K buffer power can be directly obtained from the K quantity–intensity relationship. The effective diffusion coefficient depends on, among other factors, the buffer power. Plants feed not only from exchangeable K^+ but also from nonexchangeable K^+, which mainly consists of K^+ trapped in the interlayers of nonexpanded 2:1 clay minerals (Hoagland and Martin 1933; Schachtschabel 1937).

Exchangeable K comprises that which can be exchanged with the NH_4^+ ion, and is primarily planar K^+, the interlayer K^+ of expanded 2:1 clay minerals and some of the K^+ at the interlayer edges of nonexpanded 2:1 clay minerals. Interlayer K^+ of nonexpanded clay minerals such as illites and interlayer and lattice K^+ micas (present in feldspars) constitutes the non-exchangeable K^+. The interlayer K^{++} is of particular importance in the nutrition of deep-rooted and perennial crops such as cardamom, as demonstrated by the recent work of Nair et al. (1997), and also for annual crops such as rye grass (*Lolium perenne* cv. Taptoe) (Mengel and Uhlenbecker 1993).

In most of the soil tests for K^+ availability, nonexchangeable K is not considered. For cereals such as wheat (*Triticum aestivum*) 80% of the K^+ extracted by the crop came from the nonexchangeable K pool. This is one of the most important reasons for the poor soil test crop response relationship with regard to K fertilizer application based on such tests (Kuhlmann and Wehrmann 1984). The contribution of nonexchangeable K to plant availability was assessed by 1 M HCl extraction by Schachtschabel (1961), similar to the 1 M HNO_3 extraction proposed by Pratt (1965) and McLean and Watson (1985). However, the efficiency of 1 M HCl extraction to quantify plant available K from the nonexchangeable pool has been disputed (Boguslawski and Lach 1971; Grimme 1974; Kuhlmann and Wehrmann 1984). Soils containing primarily 2:1 clay minerals such as vermiculite and illite have interlayer K in excess of crop demand. However, the availability of interlayer K of nonexpanded minerals is independent of the quantity of interlayer K as such, but dependent on its release rate depending on the type of K^+ bearing minerals (Sparks, 1987). Release of K from interlayer positions is an exchange and diffusion process (von Reichenbach, 1972). While exchange depends on the cation species and their concentration near the surface of the mineral, diffusion depends largely on the expansion of the mineral and therefore on soil moisture. Net release of K will only occur if the K concentration of the adjacent solution is low (Mengel 1985). Martin and Sparks (1983), while studying the release of nonexchangeable K^+ from sandy loam and loamy sand extracted with a H^+ charged ion exchanger resin, found large release of K^+ with a K^+ concentration of about 1–2 μ M in the contact solution. This concentration may approximate the rhizosphere concentration level. Under submerged conditions as in rice, there can be a depletion zone for K in the rhizosphere (Xu and Liu 1983). Plant roots act as a sink for K and maintain the K solution concentration at very low levels (Kuchenbuch and Jungk 1984); this would cause further release of interlayer K (Mengel 1985). These considerations point to the important fact that a precise quantification of K availability, where non-exchangeable interlayer K is concerned, hinges primarily on its release rate, which the K buffer power attempts to quantify, as we shall see in the following discussion.

Nair et al. (1997) selected cardamom (*E. cardamomum* M.) to demonstrate the importance of nonexchangeable and interlayer K on K availability *vis-a-vis* the K buffer power. The K buffer power curves were constructed by a two-step extraction in which 1 N HNO_3 was used to determine the quantity of K (Wood and De Turk 1941) and 1 N NH_4OAc was used to determine K intensity. The NH_4OAc extractant is universally used to determine exchangeable K. The contribution of nonexchangeable K to plant availability has been assessed both by extraction with 1 M HCl (Schachtschabel 1961) and 1 M HNO_3 (McLean and Watson 1985; Pratt 1965). Nair et al. (1997) regressed 1 N HNO_3 extractable K (y) over 1 N NH_4OAc extractable K (x) to obtain the K buffer power (Table 4.2 and Photo 4.1).

Data in Table 4.2 clearly indicate that the Coorg soils, which had a much higher K buffer power, produced cardamom yield which was twice that obtained in the Idukki soils. The higher K buffer power of the Coorg soils was clearly reflected in

Table 4.2 K buffer power of cardamom-growing soils from two regions of southern India extensively growing this crop

Region	Regression function (Y = a + bx)	r	Crop yield (kg/ha)
Coor g	142. 38 + 1.44 43x	0.85 61*	155
Idu kki	592. 46 + 0.91 72x	0.57 99**	80

After Nair et al. (1997)

Notes: b values in the regression function represent the K buffer power of the soil. The K buffer power refers to pooled values of soil samples obtained from 94 locations covering an area of more than 20,000 ha in the two cardamom-growing regions of southern India, namely, Coorg and Idukki. Yield data refer to the same locations

*Significant at a confidence level of 1%

**Significant at a confidence level of 5%

Photo 4.1 Base of the cardamom plant (India)

the cardamom yield. By comparison, the 1 N NH_4OAc extractable K had no significant relationship with leaf K (Table 4.3) and further the integration of K buffer power with the routine soil test K data (NH_4OAc extractable K) remarkably improved this relationship.

Cardamom is a heavy feeder of K, and in India, which grows most of this valuable spice crop, and in other countries on the Asian and African continents where this crop is grown, its K fertilizer needs are almost always based on the exchangeable K determined by 1 N NH_4OAc. Data in Table 4.3 unequivocally show the inability of this extraction to precisely predict K availability; further, data in Table 4.4 show how the situation is remarkably improved by the integration of the K buffer power into the computations. A substantial variation (302.7%) in leaf K is attributable to the K buffer power. These results have been obtained from a very extensive area (more than 20,000 ha), which demonstrates their significance (Photo 4.2).

Table 4.3 Correlation coefficients and regression functions for the relationship leaf K (Y) versus exchangeable K (x, NH_4OAc extraction)

Details	Regression function (Y = a + bx)	Correlation coefficient (r)
Leaf versus exchangeable K	Coorg Y = 1.2701 + 0.0004	0.2064
Leaf versus exchangeable K	Idukki Y = 1.644 8 + 0.0000 06	−0.006

Notes: The correlation coefficients refer to the leaf samples from 94 locations from which soil samples were also obtained to calculate the K buffer power. In cardamom, the fifth pair of leaves from the top of each panicle bearing tillers are sampled for K analysis
After Nair et al. (1997)

Table 4.4 Correlation coefficients (r) for the relationship between leaf K (Y) and exchangeable K (x, NH_4OAc extraction) for the pooled data (94 locations) from two regions (Coorg and Idukki) without (A) and with (B) K buffer power integration

	Correlation coefficient	
Details	A	B
Leaf versus exchangeable K	0.2510	0.4367**

After Nair et al. (1997)
**Significant at 1% confidence level

Photo 4.2 Effect of K fertilizer application in cardamom based on K buffer power of the soil in Karnataka State, India

The K buffer power in this instance integrates both exchangeable (NH_4OAc extraction) and nonexchangeable and/or interlayer K (HNO_3 extraction) and this gives an accurate measurement of K depletion around the plant roots. In a recent study (Mengel and Uhlenbecker 1993) on K availability from interlayer K to rye grass (*L. perenne* L.cv. Taptoe) it was observed that the rate constant (b values) obtained by correlating K released (from the interlayers of clay minerals) and time periods by a modified EUF technique was closely related to K uptake and represented the K availability index from nonexchangeable K. These rate constants, according to the authors, are of the utmost importance because they provide information on the availability of nonexchangeable K in attaining maximum yield; and a set of "critical b" values toward attaining this objective have been reported. It appears that the rate constants of Mengel and Uhlenbecker (1993) are analogous to the K buffer power values reported by Nair et al. (1997), because, though the techniques differ in their details, they have accomplished the same objective of precisely predicting K availability from the nonexchangeable pool and/or the interlayer K. The capability of tapping interlayer K varies among plant species. For instance, Steffens and Mengel (1979) found that rye grass (*L. perenne*) could feed from interlayer K^{b} for a longer period without yield depression, while red clover (*Trifolium pratense*) could not. These authors reported that since *L. perenne* had a longer and deeper root system compared to *T. pratense*, the former could grow satisfactorily while relatively low K^{b} concentration at which the latter would already suffer from K deficiency (Steffens and Mengel 1981). The differences in root mass, root length, and root morphology between monocots and dicots explain the better K^+ feeding capacity from interlayer K^+ of the former compared to the latter (Mengel 1985).

Cotton, *Gossypium hirsutum* L., is another deep-rooted long-duration crop, on which the K buffer power exerts considerable influence on K acquisition. Brouder and Cassman (1994) evaluated K uptake by cotton in a vermiculitic soil using mechanistic models and observed that initial model output produced both substantial under- and overpredictions of whole-plant K accumulation. Model predictions were greatly enhanced by estimating the K buffer power. They further concluded that the contribution of the fixed K pool to the plant available K pool was likely to be substantial and that this influence must be captured in estimates of the soil K buffer power. These studies were conducted after observing in a San Joaquin Valley cotton field in California that cotton exhibited late season K deficiency while other crop species remained unaffected. In such cases, the precise estimation of K buffer power will lead to far more dependable K fertilizer recommendations than estimations by the routine NH_4OAc extraction.

Though it has long been recognized (Schachtschabel 1937) that the soil K fraction which is not exchangeable by NH4 ions (nonexchangeable K) may be important for the supply of K to plants, it is only of late that researchers have paid more attention to this aspect. The work of Sparks and Huang (1985) has very carefully examined the release mechanism from the nonexchangeable source and the factors controlling it. Considerable portions of initially non-exchangeable K can be utilized by plant roots even within a few days (Kuchenbuch and Jungk 1984). The depletion zone, however, extends into the ambient soil for 2 mm only. Hinsinger et al. (1992) embedded phlogopite in agar and observed that the interlayer K of this mineral was entirely lost in the close vicinity of ryegrass roots within 4 days. Since the process limiting the rate of K uptake in the rhizosphere may be K transport through the soil rather than the release from minerals as such, some researchers have focused their attention on this aspect. One such example is the mechanistic mathematical model of Claassen and Barber (1976). Claassen et al. (1986) and Claassen (1990) have successfully applied the model referred to above to predict K depletion profiles in soil around plant roots. Quite recently, Meyer and Jungk (1993) have used these models to predict K uptake by test plants from exchangeable and non-exchangeable K sources. They reported that 64–79% of the K taken up by wheat (T. aestivum L.) and sugar beet (Beta vulgaris L.) was derived from the rapidly released exchangeable and 21–36% from the nonexchangeable or less mobile soil K fraction.

The buffer power describes the relationship between adsorbed K and the K concentration of the ambient solution. In simulation models it is assumed that this relationship is linear and hence independent of soil solution concentration. However, in desorption studies with soil a sharply curved buffer relationship could be found and Meyer and Jungk (1993) have referred to it. Very near the plant roots the soil can be subjected to a curved buffer function since plant roots strongly reduce the soil solution concentration.

An important aspect to be considered in the utilization of non-exchangeable K is the role of plant roots. Plant species differ in their ability to utilize nonexchangeable K and this has been attributed to the differences in root length (Mengel and Steffens 1985). Radial distance between two single roots decreases, consequently increasing root density and this would result in the overlapping of the depleted soil volumes between these roots.

This would also lead to a decrease in rate of K uptake per unit root. In the case of the rapidly diffusing K fraction, which has a higher mobility, the competition effect between roots could be very intense. There is evidence to support this view, as shown by the work of Mitsios and Rowell (1987), who observed that the contribution of nonexchangeable K increased with a corresponding increase in root density. Additionally, the differences in root hair length and density among plant species (Föhse et al. 1991) affect their ability to acquire soil K. Accordingly, the work of Meyer and Jungk (1993) has shown that K uptake was higher when they included root hairs as well in their model calculations. Since root hairs contribute to an increase in root absorbing surface, a reduction in the distance of diffusion from the site of K release to the site of K uptake, and an increase in the K concentration gradient, they can be expected to exert a pronounced effect on K availability from the less mobile K fraction.

4.2.3 The Commercial Significance of K Buffer Power Determination in K Fertilizer Management for Perennial Crops

The commercial significance of K buffer power determination for dependable K fertilizer recommendations assumes great importance in those countries which are faced with importing these fertilizers at a huge cost to the national exchequer. India is a case in point. The recent decontrol of prices of both phosphate and potassium fertilizers by the government of India resulted in an overnight escalation of their market prices. In a situation like that, the farmers become extremely wary of their field use and unless the fertilizer application is cost-effective, faith in their use, especially those mentioned above, would be shattered.

The K fertilizer recommendation for cardamom has been based exclusively on NH_4OAc extraction. The investigation of Nair et al. (1997) showed its ineffectiveness. Though the importance of K buffer power in predicting K availability has been reported earlier, these research reports related mainly to annual crops such as white clover (During and Duganzich 1979) and rye grass (Mengel and Busch 1982); the work of Nair et al. (1997) was the first of its kind in perennial crops.

4.3 The Role of the NH^+ on K^+ Availability

One of the frequent assumptions made in predicting K availability in soils is that results from binary (two-ion) exchange systems can be extrapolated to ternary (three-ion) systems using appropriate equations. The K–Ca exchange reactions in soils are often investigated in laboratory studies. Most of the research carried out on soil clay minerals and soils as exchanger surfaces (Argersinger et al. 1950; Gapon 1933; Jardine and Sparks 1984; Sposito 1981a, b; Sposito et al. 1981, 1983; Vanselow 1932) are binary exchange systems. However, field soils are at least ternary systems (Adams 1971; Curtin and Smillie 1983). The evaluation of soils as binary systems implies that these reactions can be used to predict results in ternary systems such as field soils. For this assumption to be valid, the binary exchange selectivity coefficients need to be independent of exchanger-phase composition (Lumbanraja and Evangelou 1992). But, the work of Shu-Yan and Sposito (1981) showed that it is impossible to predict exchange phase–solution phase interactions in a ternary system such as the field soil from a binary system such as the laboratory sample. This focuses the importance of the ternary systems. As far as K availability is concerned, it would be important to include the NH_4^+ ion as well. The work of Lumbanraja and Evangelou (1992) has shown that K^+ adsorption to soil surfaces is suppressed in the presence of added NH_4^+, while the adsorption of NH_4^+ to the same surface is enhanced in the presence of added K^+. These observations point to the influence of added NH_4^+ on the desorption potential (chemical potential) of adsorbed K or vice versa (Lumbanraja and Evangelou 1992) and would be relevant to the determination of K buffer power especially when agents containing NH_4^+ ions, such as NH_4OAc, are used in determining K buffer power (Nair et al. 1997). The work of Lumbanraja and Evangelou (1992), though, clearly demonstrates the effect of the NH_4^+ ion on K^+ desorption, with an increase in K^+ desorption in the presence of added NH_4^+. In its absence, it might be safe to conclude that the shape of the K buffer power curve will not appreciably change even if larger quantities of K are removed due to cropping and therefore can be considered as a relatively constant property of soils. There is evidence to support this view. These authors, while investigating the Q/I relationship on K uptake by wheat (*T. aestivum*) in calcareous vertisols and inceptisols of southwestern Spain, found that 80% of the K extracted came from the nonexchangeable K pool.

These observations, coupled with the ones discussed earlier, suggest that the precision of predicting K availability can be substantially enhanced by first quantifying the K buffer power of the soil in which the crop is intended to be grown. Admittedly, the rate-limiting steps involved in K dynamics are not entirely understood (Sparks 1987). Notwithstanding this limitation, if one must move forward in devising better management of K fertilizers in crop production, a starting point has to be made with regard to precisely quantifying K availability. Quantifying the K buffer power of soils and basing K fertilizer recommendations on this seems to be the best starting point.

4.4 K Buffer Power Measurement Versus Contemporary Soil Test K Data for Dependability of K Availability Prediction—The Context of MYR and MEY Approaches

The importance of K buffer power determination in devising effective K fertilizer recommendations must be viewed against the twin objectives of maximizing fertilizer use efficiency to attain high crop productivity—the crucial concern of developing countries—and minimizing nutrient loss to the environment—the crucial concern of industrialized countries. Potassium losses are primarily due to leaching and occur on coarse-textured soils under humid conditions (Mengel 1985). Where such leaching losses are acute, split application of K fertilizer adjusted to the crop demands and to the fluctuating weather conditions are considered a useful measure to enhance efficiency of K use by crops. Despite the fact that predictive kinetic models of K dynamics are absent and the K rate-limiting steps are still inadequately understood (Sparks 1987) (a better understanding would doubtless result in better prediction of the fate of applied K and, consequently, assist in enhancing crop yields and K uptake efficiency, through more effective K fertilizer recommendations), the K fertilizer recommendations still continue to revolve around quantifying available K, mostly by standard extractions with NH_4OAc or acids such as 1 N or 1 M HNO_3 or 1 M HCl. A case in point is the MYR (maximum yield research) or MEY (maximum economic yield) approach. The recent findings of Narang et al. (1994) in India indicate that in a rice–wheat rotation of five crop seasons, the decline in available K was 12.3 mg/g at 120–30–30/120–60–30 (N–P–K rates) and 15.6 mg/g at 180–30–30/180–60–30 (N–P–K rates) over their respective initial levels, pointing out the anomalous and inconsistent fact that at identical K application rates the reduction in available K is variable. Obviously, such a conclusion is subject to scrutiny from the viewpoint both of quantifying available K (in this case the standard NH_4OAc extraction was employed) and of the varying N and P rates where interaction between the three nutrients (N, P, and K) is bound to play a role in their individual availability. However, one cannot ignore the inadequacy of such approaches, because these contemporary soil tests fail to detect precisely the differences in soil K availability. Contradictory results have been obtained where higher K fertilizer application resulted in a corresponding higher available K (Narang and Bhandari 1990; Roy et al. 1990; Singh and Nambiar 1986) while the same practice resulted in a decline of extractable K (Hegde and Dwivedi 1992; Swarup and Singh 1987). Obviously, these inconsistent results cast a deep shadow of doubt on the dependability of contemporary soil tests for K availability in soils. A similar situation prevails in the case of rice production for MEY in Costa Rica. Using a modified Olsen's extractant, a mixture of $NaHCO3^{+}$ EDTA, Cordero and Espinosa (1994) classified soil K critical levels as low (<0.1 cmolc/l), medium (0.11–0.2 cmolc/l), and high (>0.2 cmolc/l). The K critical level was set at 0.1 cmolc/l and the K fertilizer recommendations varied from 30 to 60 kg K2O/ha. But responses to K fertilizer up to 80 kg K2O/ha were obtained even in soils with a K critical level of 0.15 cmolc/l in vertisols and mollisols with high Ca þ Mg contents (Cordero and Espinosa 1994) indicating that the K critical level based on the soil extraction is not a dependable index of availability and the situation can vary from one soil to another. The above situation would call for different K fertilizer recommendations for different soils based on extractable K and the Ca + Mg/K ratio. This disallows K fertilizer recommendations in a universal manner. These are not the limitations which the K buffer power approach suffer from. Since the K buffer power quantifies the average K concentration on the plant's root surface, though indirectly (Nair 1984a, b), it would provide greater precision in K availability prediction and, hence, result in more dependable and uniform K fertilizer recommendations. The supportive evidence provided by During and Duganzich (1979), Mengel and Busch (1982), Mengel and Uhlenbecker (1993), and Nair et al. (1997) reinforces this assumption. The MYR approach focuses on the BMP (best management practice) and aims to maximize crop yields wherever possible. Yet some of the key inputs, such as decisions on fertilizer rates, are still based on contemporary soil tests based on philosophies and procedures developed several decades ago with essentially no change in the general approach. In an interesting paper entitled "Maximum Yield Research: Friend of the Environment," presented at the 15th World Congress of Soil Science in Acapulco, Mexico, makes a strong case for this approach in maximizing crop production. He cites Borlaug (1990) the Nobel Peace Prize Winner, and Borlaug and Dowswell (1993) who said,

> The only way for agriculture to produce sufficient food to keep pace with population and to alleviate the hunger of the world's poor is to increase the intensity of agricultural production under those ecological conditions which lend themselves to intensification while decreasing the intensity of production in the more fragile ecologies. Most of the yield increase in food production needed over the next several generations must be achieved through yield increases on land now under cultivation. Moreover, these yield increases must be achieved through the application of technology already available or well advanced in the research pipeline. This will not only lead to economic development but it will also do much to solve the serious environmental problems that come as a consequence of trying to cultivate lands that are not suited for crop production. Fortunately many of the more-favored agricultural lands currently under cultivation are still producing food at yield levels far below their potential.

He further cites Cassman (1990) who mentions that the greatest challenge to global food output is

> to understand how soil fertility management affects soil quality, and to apply this knowledge to integrate all aspects of crop management to optimize yield and input utilization efficiency over the long term. Soil fertility management governs plant nutrition, which in turn influences susceptibility to disease, plant growth and yield.

In a nutshell, the MYR approach lays great stress on nutrition, yet the nutrient management aspects move along the beaten track. Unless imaginative attempts are made to revise the basic concepts of quantifying nutrient availability, even the MYR approach is bound to falter as has been clearly shown in the case of K recommendations for rice–wheat rotations in India and rice production in Costa Rica. These are but a few examples which have been reported and are reviewed here, and yet there could be many more which are not reported and which are outside the purview of this review. These facts call for a new direction in soil testing for nutrient availability. The Buffer Power Concept appears to provide this. The field experience of this new approach is detailed in the book "The Spirit of Enterprise—The 1993 Rolex Awards" which summarizes the work of Nair (1992a, b) in the farmers' fields in the Republic of Cameroon.

4.4.1 K Buffer Power and Q/I Relationship

The use of the Q/I relationship to describe the K status of soils is based on Woodruff's (1955a, b) observations that K availability to plants can be characterized by the free energy of K–Ca exchange with regard to solid–solution reactions. Beckett (1971) listed a number of assumptions as prerequisites for Woodruff's observation to be valid. An important assumption is that the rate of K uptake by plant root must be regulated by "the difference in the equivalent free energies of K and Ca as offered to the uptake sites." Maas (1969) has provided experimental evidence to substantiate this from uptake data of excised maize roots. These data demonstrate that Ca^{2+} uptake in the presence of K^+ is a competitive process. The rate of K uptake was maximized in the DG range predicted by Woodruff (1955b). Such experimental evidence of K uptake by excised maize roots provides supportive proof to the validity of Woodruff's observation that K availability could be described by the free energy of K–Ca exchange as long as all other soil factors related to plant growth are not limiting and are held constant (Beckett 1972). The use of Q/I in predicting K availability to plants in soils has been widely tested (Beckett 1972; Bertsch and Thomas 1985). However, it has been shown that the Q/I relationship does not enjoy universal acceptance since a single relationship for all soils between K uptake by a given crop and AR_K does not exist, perhaps due to the nature of the soil components regulating AR_K (Evangelou et al. 1994) (Fig. 4.5). The reader is advised to read the excellent review of Evangelou et al. (1994) for a better appreciation of the new developments and perspectives on soil K Q/I relationships. The reasons for the

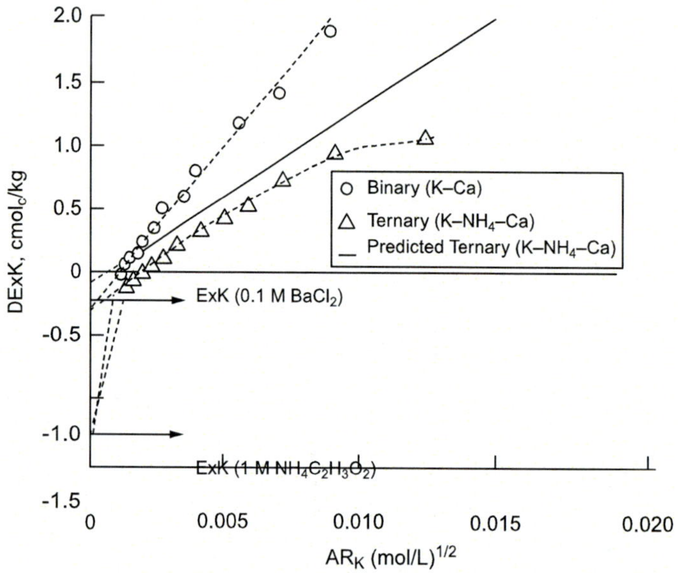

Fig. 4.5 Potassium quantity/intensity (Q/I) plots of Eden soil in the absence (binary) and the presence (ternary) of NH_4 [ternary prediction is from experimental binary Vanselow exchange selectivity coefficients for K–Ca and NH_4–Ca exchange, and from experimental ternary activity ratios (AR) for K and NH_4; ExK is exchangeable K]. (After Lumbanraja and Evangelou (1990))

failure of the Q/I concept to precisely predict soil K availability in all soils are varied and an important aspect that has to be recognized is that Q/I concept does not deal with physiological effects, but appears to be influenced by them (Evangelou et al., 1994). Mengel and Busch (1982) conducted an interesting investigation dealing with some of these physiological effects. They demonstrated that a very close correlation ($r^2 = 0.908$), reflected by a parabola ($y = c/x$; Figs. 4.6 and 4.7), existed between K buffer power and the critical K^+ concentration of the soil solution. They estimated the K buffer power by correlating exchangeable K (0.2 N BaC12 exchangeable) as y (K, quantity) with soil solution K (electro-ultrafilterable) as x (K, intensity), and the b values in the regression functions were taken as a measure of the K buffer power. Nine central European soils of extremely differing natures were tested for K availability to a test crop of *Lolium multiflorum* (cv. Avance). These authors reported that as the K buffer power increased, the critical K concentration (the critical K concentration defined as that level below which a yield increase is obtained when K fertilizer is applied) decreased.

Though the Q/I concept does not deal with physiological effects, the latter influences it. For a better appreciation of the Q/I concept vis-a-vis the K buffering effect on prediction of K uptake, one should be quite familiar with plant nutrition concepts. Even if nutritional effects are assumed to play only a limited role in K uptake under a well-defined experimental setup, the ability of a soil to replenish K^+ in the soil solution must be considered a highly complex process and not solely dependent

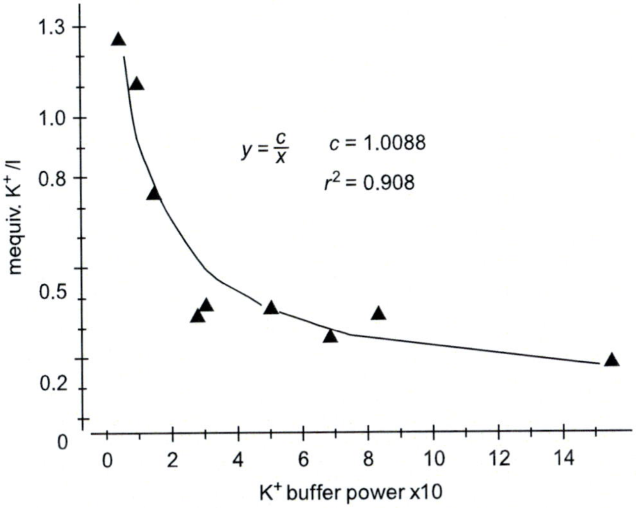

Fig. 4.6 Relationship between the critical K^+ concentration of the soil solution and the K^+ buffer power: y = critical K^+ concentration; x = K^+ buffer power. (After Mengel and Busch (1982))

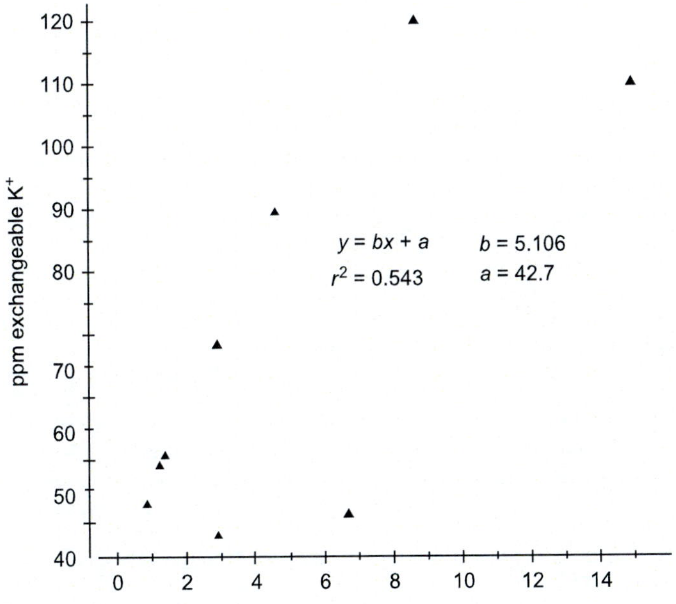

Fig. 4.7 Relationship between the critical level of exchangeable K^+ and the K^+ buffer power: y = critical level of exchangeable K^+; x = K^+ buffer power. (After Mengel and Busch (1982))

on K–(Ca + Mg) exchange alone (Evangelou et al. 1994). K availability is also controlled by additional factors, such as hysteresis or exchange irreversibility effects, anion effects, multiplication effects, potential-determining ion effects, and kinetic effects (Evangelou et al. 1994). To predict K availability with precision, these factors need to be further scrutinized. But, most importantly, it is the soil/soil solution–plant root surface interactions which merit closest scrutiny in predicting K availability. The importance of K buffer power has to be examined in this context. Since the ultimate objective in perfecting prescriptive soil management practices with regard to K fertilizer application, vis-a-vis soil testing for plant available K, hinges on the soil/soil solution–plant root interactive effects, and also since plant root surface K concentration during the dynamic state of plant growth is nearly impossible to measure in a field soil, Nair (1984a, b) has argued that a precise quantification of K buffer is an important prerequisite to accurately predict K availability and must be considered as such as an indirect measure of the plant root surface K concentration. The experimental evidence provided by Mengel and Busch (1982) and Nair et al. (1997) substantiates this view.

References

Adams F (1971) Ionic concentrations and activities in soil solutions. Soil Sci Soc Am Proc 35:420–426

Argersinger WJ Jr, Davidson AW, Bonner OD (1950) Thermodynamics and ion exchange phenomena. Trans Kans Acad Sci 53:404–410

Barekzai A (1984) Alterung von wasserlo¨ slichem Phosphat-untersucht in Giefa¨ß und Modellversuchen. PhD thesis. FB 19, Justus Liebig University, Giessen

Barrow NJ (1983) On the reversibility of phosphate sorption by soils. J Soil Sci 34:751–758

Beckett PHT (1971) Potassium potentials—a review. In: Potash review. International Potash Institute, Bern, pp 1–41

Beckett PHT (1972) Critical cation activity ratios. Adv Agron 24:376–412

Beckwith RS (1965) Sorbed phosphate at standard supernatant concentration as an estimate of the phosphate needs of soils. Aust J Exp Agric Anim Husb 5:52–58

Bertsch PM, Thomas GW (1985) Potassium status of temperate region soils. In: Munson RD (ed) Potassium in agriculture. American Society of Agronomy, Madison, pp 131–162

Boguslawski EV, Lach G (1971) Die K-Nachlieferung des Bodens im Pflanzenexperiment im Vergleich mit dem austauschbaren Kalium. Z. Acker Pflanzenbau 134:135–164

Borlaug NE (1990) The challenge of feeding 8 billion people. Farm Chem Int (Summer):10–12

Borlaug NE, Dowswell CR (1993) Fertilizer: to nourish infertile soil that feeds a fertile population that crowds a fragile world. Better Crops Plant Food 77(3):6–7

Bowman RA, Olsen SR (1985a) Assessment of phosphate buffering capacity. 1. Laboratory methods. Soil Sci 140:287–291

Bowman RA, Olsen SR (1985b) Assessment of phosphate buffering capacity. 2. Green-house methods. Soil Sci 140:387–392

Brouder SM, Cassman KG (1994) Evaluation of a mechanistic model of potassium uptake by cotton in vermiculitic soil. Soil Sci Soc Am J 58:1174–1183

Cassman KG (1990) The role of soil fertility research in developing sustainable food production systems. Better Crops Plant Food 74(4):16–19

Claassen N (1990) Die Aufnahme von Na¨hrstoffen aus dem Boden durch die ho¨ here Pflanze als Ergebnis von Verfu¨ gbarkeit und Aneignungsvermo¨ gen. Severin-Verlag, Gottingen
Claassen N, Barber SA (1976) Simulation model for nutrient uptake from soil by growing plant root system. Agron J 68:961–964
Claassen N, Syring KM, Jungk A (1986) Verification of a mathematical model by simulating potassium uptake from soil. Plant Soil 95:209–220
Cordero A, Espinosa J, (1994) Maximum economic rice yield in Costa Rica. Research for maximum yield in harmony with nature. In: XV World Congress of Soil Science, Acapulco, Mexico, July 10–16, pp 29–39
Curtin D, Smillie GW (1983) Soil solution composition as affected by timing and incubation. Soil Sci Soc Am J 47:701–707
During C, Duganzich DM (1979) Simple empirical intensity and buffering capacity measurements to predict potassium uptake by white clover. Plant Soil 51:167–176
Evangelou VP, Wang J, Phillipis RE (1994) New developments and perspectives on soil potassium quantity/intensity relationships. Adv Agron 52:173–227
Fohse D, Claassen N, Jungk A (1991) Phosphorus efficiency of plants. II. Significance of root radius, root hairs, and cation–anion balance for phosphorus influx in seven plant species. Plant Soil 132:261–272
Fox RL (1981) External phosphorus requirements of crops. In: Dowdy RH et al (eds) Chemistry in the soil environment. ASA Spec. Publ. 40 ASA and SSSA, Madison, pp 223–239
Fox RL, Kamprath EJ (1970) Phosphate sorption isotherms for evaluating the phosphate requirement of soils. Soil Sci Soc Am Proc 34:902–907
Gapon EN (1933) Theory of exchangeable adsorption in soil. J Gen Chem (USSR) 3, 144–163 (Abstract in Chem Abs 28, 41–49, 1934)
Grimme H (1974) Potassium release in relation to crop production. In: International Potash Institute (ed) Potassium research and agricultural production. International Potash Institute, Berne, pp 131–136
Hegde DM, Dwivedi SS (1992) Nutrient management in rice–wheat cropping system in India. Fertil News 37:27–42
Hinsinger P, Jaillard B, Dufey JE (1992) Rapid weathering of a trioctahedral mica by roots of ryegrass. Soil Sci Soc Am J 56:977–982
Hoagland R, Martin JC (1933) Absorption of potassium by plants in relation to replaceable, non-replaceable and soil solution potassium. Soil Sci 36:1–34
Holford ICR (1988) Buffering of phosphate in the soil solution during growth of ryegrass compared with buffering by sorption. Plant Soil 111:3–9
Jardine PM, Sparks DL (1984) Potassium–calcium exchange in a multi-reactive soil sys-tem II. Thermodynamics. Soil Sci Soc Am J 48:45–50
Kuchenbuch R, Jungk A (1984) Wirkung der Kaliumdu¨ ngung auf die Kalium verfu¨gbarkeit in der Rhizospha¨re von Raps. Z. Pflanzenerna¨hr. Bodenkd 147:435–448
Kuhlmann H, Wehrmann J (1984) Kali-Du¨ ngeempfehlung auf der Grundlage von 81K- Du¨ ngeversuchen zu Gertreide und Zuckerru¨ ben. Z. Pflanzenerna¨hr. Bodenkd. 147
Larsen S (1967) Soil phosphorus. Adv Agron 19:151–210
Lumbanraja J, Evangelou VP (1990) Binary and ternary exchange behavior of potassium and ammonium at low exchangeable fractional loads. Soil Sci Soc Am J 54:698–705
Lumbanraja J, Evangelou VP (1992) Potassium quantity–intensity relationships in the presence and absence of NH4 for three Kentucky soils. Soil Sci 154:366–376
Maas EV (1969) Calcium uptake by excised maize roots and interactions with alkali cations. Plant Physiol 44:985–989
Martin HW, Sparks DL (1983) Kinetics of nonexchangeable potassium release from two coastal plain soils. Soil Sci Soc Am J 47:883–887
Mattingly CEG, Russell RD, Jephcott BM (1963) Experiments on cumulative dressings of fertilizers on calcareous soils in Southwest England. II Phosphorus uptake by rye-grass in the green house. J Sci Food Agric 14:629–637

McLean EO, Watson ME (1985) Soil measurement of plant-available potassium. In: Munson RD (ed) Potassium in agriculture. ASA, CSSA, and SSA, Madison, pp 277–308
Mengel K (1985) Dynamics and availability of major nutrients in soils. Adv Soil Sci 2:65–131
Mengel K, Busch R (1982) The importance of the potassium buffer power on the critical potassium level in soils. Soil Sci 133:27–32
Mengel K, Kirby EA (1980) Potassium in crop production. Adv Agron 33:59–110
Mengel K, Steffens D (1985) Potassium uptake of ryegrass (Lolium perenne) and red clover (Trifolium pratense) as related to root parameters. Biol Fertil Soils 1:53–58
Mengel K, Uhlenbecker K (1993) Determination of available interlayer potassium and its uptake by ryegrass. Soil Sci Soc Am J 57:761–766
Meyer D, Jungk A (1993) A new approach to quantify the utilization of nonexchangeable soil potassium by plants. Plant Soil 149:235–243
Mitsios IK, Rowell DL (1987) Plant uptake from exchangeable and non-exchangeable potassium. I Measuring and modelling for onion roots in a chalky boulder clay soil. J Soil Sci 38:53–63
Nair KPP (1984a) Towards a better approach to soil testing based on the buffer power concept. Proc 6th Int Colloquium for the Optimization of Plant Nutrition, 2–8 September, Pierre-Martin Prevel, Montpellier, France 4:1221–1228
Nair KPP (1984b) Zinc buffer power as an important criterion for a dependable assessment of plant uptake. Plant Soil 81:209–215
Nair KPP (1992a) Measuring P buffer power to improve routine soil testing for phosphate. Eur J Agron 1(2):79–84
Nair KPP (1992b) Soil testing procedure to economise on fertilizer use in land management. In: The spirit of enterprise—the 1993 Rolex awards. The Rolex Foundation, Geneva, pp 323–325
Nair KPP, Mengel K (1984) Importance of phosphate buffer power for phosphate uptake by rye. Soil Sci Soc Am J 48:92–95
Nair KPP, Sadanandan AK, Hamza S, Abraham J (1997) The importance of Potassium Buffer Power in the Growth and Yield of Cardamom. J Plant Nutr 20(7 & 8):87–997
Narang RS, Bhandari AL (1990) Integrated nitrogen management in rice–wheat system Paper presented at the Rice–Wheat Workshop. Directorate for Cropping System in Punjab, India. Proceedings of Symposium on Maximum Yield Research, November 16–18, New Delhi, India
Narang RS, Tiwanria US Dev G (1994) Maximum yield research studies in rice–wheat system and soil productivity—the Indian experience. In: Research for Maximum Yield in Harmony with Nature. Satellite Symposium Sponsored by PPI. XV World Congress of Soil Science, Acapulco, Mexico, July 10–16, pp 46–55
Nemeth K (1979) The availability of nutrients in the soil as determined by elec-troultrafiltration (EUF). Adv Agron 31:155–188
Okajima H, Kubota H, Sakuma T (1983) Hysteresis in the phosphorus sorption and desorption process of soils. Soil Sci Plant Nutr 29:271–283
Olsen SR, Cole CV, Watanabe FS, Dean LA (1954) Estimation of available phosphorus in soils by extraction with sodium bicarbonate. USDA Cir. No. 939
Pagel H, van Huay H (1976) Wichtige Parameter der Phosphat-Sorptionskurven einiger Bo¨ den der Tropen und Subtropen und ihre Zeitliche Vera¨nderung durch P—Du¨ ngung. Arch Acker Pflanzenbau Bodenkd 20:765–778
Pratt PF (1965) Potassium. In: Black CA et al (eds) Methods of soil analysis, part 2, Agron. Monogr. 9. ASA, Madison, pp 1023–1031
Raven KP, Hossner LR (1994) Sorption and desorption—quantity–intensity parameters related to plant-available soil phosphorus. Soil Sci Soc Am J 58:405–410
Roy HK, Kumar A, Sinha K (1990) Response of wheat to K in red loam (Alfisols) soils of Ranchi. J Potassium Res 6:23–28
Schachtschabel P (1937) Aufnahme von nicht austauschbarem Kalium durch die Pflanze. Z. Pflanzenerna¨hr. Bodenkd 48:107–133

Schachtschabel P (1961) Fixierung und Nachlieferung von Kalium-und Ammonium-Ionen. Beurteilung und Bestimmung des Kaliumversorgungs-grades von Bo¨ den. Landwirtsch Forsch 14:29–47
Schofield RK (1955) Can a precise meaning be given to "available" soil phosphorus? Soil Fertil 18:373–375
Schuller H (1969) Die CAL-Methode, eine neue Methode zur Bestimmung des pflanzen verfu¨ gbaren Phosphates in Bo¨ den. Z. Pflanzenerna¨hr. Bodenkd 123:48–63
Shu-Yan C, Sposito G (1981) The thermodynamics of ternary cation exchange systems and the subregular model. Soil Sci Soc Am J 45:1084–1089
Singh GB, Nambiar KKM (1986) Crop productivity and fertility under intensive use of chemical fertilizers in long-term field experiments. Indian J Agron 31:115–127
Sparks DL (1987) Potassium dynamics in soils. Adv Soil Sci 6:2–63
Sparks DL, Huang PM (1985) Physical chemistry of soil potassium. In: Munson RD (ed) Potassium in agriculture. American Society of Agronomy, Madison, pp 201–276
Sposito G (1981a) Cation exchange in soils. A historical and theoretical perspective. In: Dowday RH (ed) Chemistry in the soil environment. Soil Science Society of America, Madison
Sposito G (1981b) The thermodynamics of the soil solution. Clarendon Press, Oxford
Sposito G, Holtzelaw KM, Johnston CT, Le Vesque CS (1981) Thermodynamics of sodium-copper exchange on Wyoming bentonite at 198 K. Soil Sci Soc Am J 45:1079–1084
Sposito G, Holtzelaw KM, Charlet L, Jonany C, Page AL (1983) Sodium-calcium and sodium magnesium exchange on Wyoming bentonite in perchlorate and chloride background ionic media. Soil Sci Soc Am J 47:51–56
Steffens D, Mengel K (1979) Das Aneignungsvermo¨ gen von Lolium perenne im Vergleich zu Trifolium pratense fu¨ r Zwischenschilht-Kalium der Touminerale. Landw Forsch SH 36:120–127
Steffens D, Mengel K (1981) Vergleichende Untersuchungen zwischen Lolium perenne und Trifolium pratense u¨ ber das Aneignungsvermo¨ gen von Kalium. Mitt Dtsch Bodenk Ges 32:375–386
Swarup A, Singh KN (1987) Movement of potassium in a sodic soil profile as influenced by long-term use of inorganic fertilizers under rice–wheat rotation. J Indian Soc Soil Sci 35:253–256
van der Paauw F (1971) An effective water extraction method for the determination of plant-available soil phosphorus. Plant Soil 34:467–481
Vanselow AP (1932) Equilibria of the base-exchange reactions of bentonites, permutities, soil colloids, and zeolites. Soil Sci 33:95–113
von Reichenbach H (1972) Factors of mica transformation. In: Potassium in soil. International Potash Institute, Berne, pp 33–42
Wood LK, De Turk EE (1941) The adsorption of potassium in soils in replaceable forms. Soil Sci Soc Am Proc 5:152–161
Woodruff CM (1955a) Ionic equilibria between clay and dilute salt solutions. Soil Sci Soc Am Proc 19:36–40
Woodruff CM (1955b) The energies of replacement of calcium by potassium in soils. Soil Sci Soc Am Proc 19:167–171
Xu ML, Liu ZY (1983) The nutrient status of soil–root interface. II Potassium accumulation and depletion in rhizosphere soils. Acta Pedol Sin 20:295–302

Chapter 5
Quantifying the Buffer Power for Precise Availability Prediction—Heavy Metals

Abstract Of all the micronutrients of importance in plant nutrition, Zinc takes place of priority, as it is involved in several life sustaining metabolic processes for humans, animals and plants. One of the most serious consequences of the highly "chemical centric" green revolution is to impoverish world soils of Zinc. Vast stretches of both Central Asian (in particular, countries like Turkey) and South Asian (in particular countries like India and Pakistan) soils are now heavily Zinc deficient. The chapter would focus on the relevance of the "**Buffer Power Concept**" in ameliorating the condition, by investigating the centrality of the concept in the nutrition of annual crops like wheat, the principal staple of these countries, and also, perennial crops like Black pepper, the economic mainstay of States like Kerala in southern India. The details of analytical techniques involved in quantifying Zinc Buffer Power are also discussed in the chapter.

5.1 Zinc

Since Zn, among the heavy metals, has become the most problematic on a global scale, the scope of this review is confined to this very important plant nutrient. There is a great paucity of published material on the effect of buffer power on availability of heavy metals.

Plants obtain most of their fertilizer Zn from reaction products and not applied sources as such, implying that any source of Zn added to soil has to necessarily conform to a chain reaction involving adsorptive, desorptive, and resorptive processes that govern the maintenance of an equilibrium between adequate Zn concentration in the soil solution nearest to the zone of Zn depletion on the one hand and plant uptake on the other. The Zn effective diffusion coefficient or Zn buffer power defines this. As Zn concentration in soil solution is normally very low, the supply to plant roots by mass flow can only account for a very small fraction of plant demand. For instance, with a transpiration coefficient of 300 l/kg dry matter and a corresponding Zn concentration of 10^{-7} M in the soil solution, approximately 2 mg of Zn can be supplied by mass flow against a demand of 10–30 mg Zn/kg dry weight of plant tissue. In calcareous soils, as the Zn concentration is of a much lower order of approximately 10^{-8} M, the supply by mass flow could be very much lower

© The Author(s), under exclusive license to Springer Nature Switzerland AG 2021

K. P. Nair, *Thermodynamics of Soil Nutrient Bioavailability*,
https://doi.org/10.1007/978-3-030-76817-1_5

(Marschner 1994) indicating that mass flow can only contribute very negligibly to meet plant needs of Zn. Hence, Zn movement to the plant root surface is principally by diffusion and is essentially confined to a zone around the plant root which hardly extends beyond the root hair cylinder (Marschner 1994).

Though published papers on the relevance of Zn buffer power in soils to Zn availability are very limited, there is no dearth on other aspects. Genotypic differences to Zn nutrition in a very important crop such as maize, with regard to its hybrids, inbreds, and composites, were reported almost two decades ago in India (Nair and Prabhat 1977); this study focused on the importance of genetic engineering for tailoring maize varieties for better efficiency of Zn utilization, while the complicated mechanism of Zn–P–Fe interaction was researched even prior to this (Nair and Babu 1975). In a recent review on the mechanism of Zn uptake, Marschner (1994) indicated that flow culture experiments with various species showed adequate ranges of Zn concentration in the range of 6×10^{-8} to 8×10^{-6} M which are concentrations greater than those that would be expected in the soil solution of most soils. He further pointed out that although work using chelate-buffered solutions has indicated adequate Zn concentrations between 10^{-10} and 10^{-11} M, extremely low adequate Zn concentrations required a concomitant excess of about 100 mM Zn-chelate as buffer at the plasma membrane of the root cells. This implies a need for an unlimited Zn pool for replenishment of Zn^{2+} at the plasma membrane. When plants grow in soil, it is impossible to expect a Zn buffer of this size to exist, and free Zn^{2+} and chelated-Zn concentrations will be at least threefold lower. Hence, critical deficiency or sufficiency concentrations obtained through research employing chelate-buffered solutions cannot be applied to soil-grown plants.

Most of the work on Zn availability to plants in soil is based on chemical extractions among, which DTPA is the most frequently used. The DTPA extraction quantifies a labile fraction of soil Zn comprising water soluble, exchangeable, adsorbed, chelated, and some occluded Zn. The critical soil level of DTPA extractable Zn can vary from 0.3 to 1.4 mg/kg soil which equates to about 900–4200 g/ha of Zn in heavy soils and about 600–2800 g/ha Zn in light soils in the plow layer (0–20 cm). The crop requirements, on the other hand, are quite small, in the range of 100–300 g/ha for a total dry matter production of about 10 t/ha (Marschner 1994). The inadequacy of DTPA extraction to reflect plant Zn demand shows that other important factors, such as replenishment of soil solution Zn (Nair 1984a, b), mobility, and transport to the root surface (Nair et al. 1984; Wilkinson et al. 1968), and also the activity of the roots themselves (Marschner 1994; Wilkinson et al. 1968), are involved. Since the Zn buffer power is intricately involved in all three factors, the focus of this review is mainly on that attribute.

As early as three decades ago it was suggested that colloidal Zn was released by some specific process associated with root activity (Wilkinson et al. 1968). Conditions in the rhizosphere and particularly root-induced changes markedly affect Zn availability. A difference in rhizosphere pH of as much as 2, higher or lower compared to bulk soil, can be expected to occur as a result of imbalance in ionic uptake. For instance, any acidifying fertilizer such $(NH_4)_2SO_4$ can result in a net excretion of H^+ ions and others such as NH_4NO_3 can result in a net excretion of

HCO_3^- or OH^- ions. Additionally, secretion of organic acids and enhanced CO_2 production, as well, will affect rhizosphere pH, and all of the above-mentioned changes will markedly affect Zn availability. However, the scope of this review is confined to the kinetic/dynamic aspects of the changes occurring in the rooting zone mirrored in the Zn buffer power rather than changes in soil reaction in the rhizosphere per se on Zn availability.

The distribution of Zn between the solid and solution phases can be described by the buffer power. The availability of soil Zn to the plant depends on the initial Zn concentration, Zn buffer power and effective diffusion coefficient (Barber 1984). The Langmuir equation gives the relation between B and C1 as

$$\frac{C_1}{x/m} = \frac{1}{aB} + \frac{C_1}{B}$$

where C1 is the Zn concentration in the soil solution, x/m is the amount of Zn adsorbed per unit of soil, B is the adsorption maximum, and a is a constant related to the soil's bonding energy for Zn. A straight line is obtained when C1/(x/m) is plotted against C1 with a slope of 1/B and intercept of 1/aB. The inverse of C1/(x/m) = b, the Zn buffer power, where C1 and x/m are both expressed in volume units (Barber 1984). Using this approach, Shuman (1975) estimated the buffer power values varying from 5 to 100 for four soils representing different major physiographic regions of Georgia. Based on the diffusion model of Drew et al. (1969), Nair et al. (1984a, b) has argued that the $\bar{c}$ in the equation $U = 2\pi\ \alpha\ a\ a^{-c}\ t$ (Drew et al. 1969), where U is the quantity of Zn absorbed per centimeter root length, a is the root radius in cm, α is the root absorbing power, c is the average Zn concentration on the root surface, and t is the duration of the absorption period, in fact represents an indirect measure of the Zn buffer power. As we already know, the bulk of Zn uptake is by diffusion (Barber 1984; Elgawhary et al. 1970; Wilkinson et al. 1968). This diffusive process will maintain a concentration gradient in the rooting zone. This concentration gradient will directly affect Zn uptake because of its effect on the average Zn concentration on the root surface. The Zn buffer power will affect this concentration gradient, because the rate of Zn depletion and/or replenishment is mirrored by it. In a sense, the effective diffusion coefficient and the buffer power are analogous for nutrients which are principally absorbed by the plant root through diffusive processes (Nair 1989). Hence, the crucial question to examine would be the role of Zn buffer power in influencing Zn availability for plant uptake.

5.2 Quantifying Zn Buffer Power

Nair et al. (1984a, b) used a novel method to quantify Zn buffer power. Nair (1990) has demonstrated the importance of Zn and P buffer power in affecting their bio availability to growing plants. Heavy metal pollution is a major environmental hazard in Europe, especially on the highways due to automobile exhausts; as a consequence, a number of soils are very highly contaminated. Such soils are unsuitable for a precise study on the impact of Zn buffer power on plant Zn uptake, and the choice of soil was narrowed to a heavy clay in the northern part of Belgium containing 1.1 ppm total Zn, 5.5 ppm DTPA extractable Zn, and 0.13 ppm electro-ultra-filterable Zn. The soil was acidic (pH 6.6, 1 N KCl), low in organic carbon (1.8%), high in $CaCO_3$ (16.5%), containing beidellite as the predominant clay mineral. Zn buffer power was quantified by regressing DTPA extractable Zn (y) on electro-ultra-filterable Zn (x) by a stepwise extraction of incubated soil, and the b value in the regression function represented the Zn buffer power. EUF has been employed for a rapid estimation of the intensity of a range of heavy metals, especially in central, western, and eastern European soils (Nemeth 1979; Nemeth and Recke 1982). However, earlier investigations of these authors did not give satisfactory results: analysis of only the alkaline cathode filtrate is insufficient to give precise information about availability indices, primarily due to the fact that the heavy metals are collected by the cathode filter as hydroxides and hydrated oxides, which are hardly soluble in the alkaline cathode filtrate. However, as these oxides comprise soluble and desorbable ions of the heavy metal in question, they are of importance to plant nutrition; Nair (1984a, b) used a modified procedure to bring the oxides into an acid medium for subsequent analysis. The studies indicated that Zn intensity alone is insufficient to precisely predict Zn uptake by maize used as a test plant in the investigation. The relevant data are given in Table 5.1.

Data in Table 5.1 show that the precision of predicting Zn concentration in the maize top increased remarkably, as substantiated by the increase in the coefficient of determination which works out to 65.9%. With regard to total Zn uptake by the maize plant, the increase is less marked, yet substantial. The data show that the major factor responsible for the variance in Zn concentration in the maize top and total Zn uptake by the maize top is the Zn buffer power, and not Zn intensity as quantified by EUF. In a separate experimental setup, Nair et al. (1984) monitored Zn transport by simulation experiments in a laboratory setup using the same field soil and obtained proof to support the hypothesis that it is essentially the buffered Zn that contributes to the major uptake process by the test plant. Dissolved salts in a field soil must move with the liquid, the net movement from a given reference plane

Table 5.1 Correlation coefficients (r) between Zn intensity and maize top Zn concentration and total Zn uptake by plant top at harvest without (A) and with (B) Zn buffer power integration

Details	Correlation coefficient	
	A	B
Zn intensity versus maize top Zn concentration	0.290	0.812*
Zn intensity versus total Zn uptake by maize top	0.356	0.618**

After Nair et al. (1984a, b)
*Significant at 0.1% confidence level
**Significant at 1% confidence level

being dependent on the activity gradient of salt and water. This naturally brings us to the important question of solute movement within the soil matrix which is dependent on the soil conductivity. Buffered Zn would be in the bulk soil solution to be further transported to plant roots. This transport, though dependent on the soil conductivity, would also be linked to the quantum of Zn per se in the bulk soil solution, which, in turn, would be dependent on the Zn buffer power of the soil in question. Nair et al. (1984) quantified these relationships. In a field soil, it is the buffered Zn which maintains a concentration gradient between the bulk soil solution and the root surface sufficiently conducive for optimal plant uptake. The investigation also showed the ineffectiveness of DTPA extraction as a reliable soil test for Zn availability. EUF is being put to use increasingly for commercial soil testing in many countries in Europe, notably, for sugar beet cultivation in Austria. Determination of Zn intensity singly may not provide reliable information regarding plant available Zn. Besides Zn intensity and Zn buffer power, other soil factors have only a minor influence on Zn absorption. EUF has the advantage of a rapid assessment of the Zn intensity factor. Combining the advantage of rapid assessment of the intensity factor of the nutrient with its corresponding buffer power might greatly enhance the reliability of fertilizer recommendations not only for Zn, but, perhaps, for other heavy metals as well, making EUF a reliable tool in soil testing for heavy metals.

5.3 Case Studies with Asian Soils

5.3.1 South Asian Soils

The 3rd International Zinc Symposium in Hyderabad, India, in which 195 delegates from 30 countries participated, focused on the cruciality of Zn in both plant and human nutrition. Zn is assuming the role of a major constraint in crop production. There is wide-spread Zn deficiency in topical soils. This section of the review is divided into two parts, where the focus will be first on (A) South Asian soils, India being the classical example and (B) Central Asian soils, Turkey, where severe Zn deficiency in the main wheat growing belt of the country (Anatolia) is threatening wheat production and adversely affecting human nutrition.

Investigations in Indian soils. The lateritic belt of Kerala State, in southern India, was chosen as the test soil. The economic mainstay of the State is black pepper (*Piper nigrum* L.). The crop is very susceptible to Zn deficiency and among the major diseases causing havoc in black pepper plantations, the onset of the "Quick Wilt" caused by *Phytophthora* spp. (*Phytophthora capsici*) has been attributed to a severe Zn deficiency in the soil. The "official" recommendation for supplemental Zn fertilizer is based on the classical DTPA extraction, and a "blanket" recommendation for the entire State is made, where the farmers are advised to apply Zn fertilizer at the rate of 25 kg/ha for the entire State. This author has, for long, suspected the advisability of making such a "blanket" recommendation, and had suggested, in its place, investigating the Zn buffering mechanism of the soils and then calibrating Zn fertilizer recommendations based on this. Zn fertilizer is quite expensive, and, at the prevalent rate of approximately Rs. 300/kg (approximately U.S. $/kg) could turn out to be prohibitively expensive for a poor black pepper farmer. In the following paragraphs, the salient results are summarized.

5.3.1.1 Quantifying the Z Buffer Power of the Pepper-Growing Soils

Nair (1984a) has used EUF to quantify Zn "intensity" in measuring Zn buffer power of European soils. Because the procedure is very highly sophisticated, and due to the non availability of such expensive instruments in a country like India, and many other developing countries, a simple adsorption–desorption equilibrium technique was developed by the author (Nair 2002) to quantify Zn buffer power of black pepper-growing soils. For this, 200 g of representative soil samples from the principal black pepper-growing regions were selected from the State, incubated at 60% of the maximum water holding—capacity with graded rates of Zn, over a fortnight, by maintaining the water regime to constancy. At the end of the period, the soil samples were extracted with 0.01 M $CaCl_2$ over a 24-h period and the extract tested for Zn using atomic absorption spectrophotometry. This represented Zn "intensity." Separate extractions were made with DTPA, which represented the Zn "quantity." Data in Table 5.2 indicate that the Zn buffer power values of the black pepper-growing soils varied widely (Photo 5.1).

Table 5.2 Zinc buffer power of black-pepper-growing soils

Soil	"r" Value	"b" Value
Peruvannamuzhi	0.8337***	0.7824
Thamarasseri	0.9304***	1.5786
Ambalavayal	0.9604***	3.0358

Notes: All the "r" values are very highly significant, at a confidence level of 99.9% (***) implying that the adsorption–desorption equilibrium technique developed by the author provides an accurate measure of the Zn buffer power. Further, the soils showed widely varying Zn buffer power, as shown by the "b" values, with the soils from *Peruvannamuzhi* region least buffered for Zn, while that from the *Ambalavayal* region most buffered, with the soils from *Thamarasseri* region intermediate in their Zn buffer power. These findings have remarkable impact, as we shall see from the results incorporated in Table 5.4

Photo 5.1 Effect of Zn fertilizer application on Bush pepper based on the soil's Zn buffer power in Kerala State, India

Data in Table 5.2 bring to light two very important facts. First, the very highly significant correlation coefficients ("r" values) prove that the technique employed allows a precise determination of the Zn buffer power of the soils investigated. Second, the soils varied widely in their Zn buffer power. It is this fact which has to be critically viewed against the existing Zn fertilizer recommendation, emanating from routine soil tests using the universally employed DTPA extraction, to understand the cruciality of the Zn buffer power for precise formulations of Zn fertilizer application for black pepper.

Data in Table 5.3 clearly indicate that integrating Zn buffer power has improved the interrelationship between DTPA routine soil test versus both Zn concentration in plant top and Zn uptake. Most remarkably, the overall relationship between DTPA soil test and dry matter production not only improved very significantly when Zn buffer power was integrated into the computations, but from a negative value to a positive one. This implies two very important conclusions. First, the universally adopted DTPA soil test for Zn is unsuitable for the soils investigated. Second, it is the Zn buffer power that is the decisive factor which determines dry matter production, implying, higher the Zn buffer power, more the yield. Results in Table 5.4 substantiate this very important conclusion.

Data in Tables 5.1, 5.5, and 5.4 have to be reviewed in conjunction to precisely understand the cruciality of the "Buffer Power Concept." Black pepper is the economic mainstay of the Kerala State. Of late, widespread Zn deficiency has been observed in the State, and the onset of the most devastating "Quick Wilt" disease and "Foot Rot" disease, both of which can exterminate a black pepper plantation in a matter of a week's time, has been attributed to the infestation by *Phytophthora* fungi, which thrives best in Zn-deficient soils. The scientists at the Indian Institute

Table 5.3 Linear regression functions (Y = a + bx) and correlation coefficients ("r" values) for the relationship between DTPA extractable Zn (Zn "quantity" Y) and 0.01 M $CaCl_2$ extractable Zn (Zn "intensity" x)

Regression function				
Soil	Adsorption–desorption equilibrium	Soil incubation	"r" value A	B
Konya (Central Anatolia)	Q = 201*bI*+ 5.3	Q = 679*bI* – 5.9	0.951*	0.978*
Koruklu (gap region)	Q = 17*bI* + 17.9	Q = 375*bI* – 8.7	0.015	0.784*
German soil (reference soil)	Q = 8*bI* + 6.5	Q = 11*bI* + 7.8	0.972*	0.942*

Notes: Q = Zn "quantity," *I* = Zn "intensity," b = Zn buffer power. A without the integration of Zn buffer power ("b" values) in the computations, B with the integration. The experimental technique employed has resulted in very highly significant correlation coefficients ("r" values), except in the case of soils from the Gap region (*Koruklu*) in the case of the incubation experiments, which suggest the high accuracy of the adsorption–desorption equilibrium technique to determine Zn intensity, which is a true reflection of the Zn buffer. Note the remarkable improvements in the "r" values (B compared to A) which unequivocally show that the integration of the Zn buffer power remarkably improves the relationship. How this affects plant uptake, and, eventually wheat yields, will be discussed in the following tables and explanation of experimental details
*Significant at a confidence level of 99.9%

Table 5.4 Black pepper yields from farmers' fields weighted against the Zn buffer power

Region	Yield (kg/vine) Targeted	Actual	Deviation (%)
Peruvannamuzhi	0.241	0.401	+66
Thamarasseri	0.490	0.487	+0.6

Notes: Target weighting was done against the highest vine yield obtained from the *Ambalavayal* region. Note the remarkable closeness between targeted and actual yields in *Thamarasseri* region, whereas in *Peruvannamuzhi* region, there is a huge deviation. Soils in the *Peruvannamuzhi* region are atypical black pepper soils. Ironically the "official" recommendation for Zn fertilizer recommendation emanate from experiments conducted in *Peruvannamuzhi*, as the black pepper research station, affiliated to the Indian Institute of Spices Research in Calicut, Kerala State, is situated here

Table 5.5 Correlation coefficients ("r" values) for the interrelationship between routine DTPA soil test versus Zn concentration, Zn uptake and dry matter production without (1) and with (2) Zn buffer power integration in black pepper

Details	1	2
Zn concentration versus DTPA test	0.88*	0.92*
Zn uptake versus DTPA test	0.78	0.86*
Dry matter production versus DTPA test	−0.74	0.78*

*Significant at 99.9% confidence level

of Spices Research in Calicut, Kerala State, India, under the administrative control of the Indian Council of Agricultural Research in New Delhi, have made a "blanket" recommendation of soil application of 23 kg $ZnSO_4$/ha, when DTPA extractable Zn is less than 1.6 ppm. In terms of monetary input, this would translate to more than U.S. $ 25/ha in terms of farmer investment. Because the soils are variously Zn buffered (Table 5.2), a single "blanket" dose as the Kerala Government is propagating is totally unscientific, because, where the farmer needs to apply only lesser quantities of Zn, he would end up applying more and thus waste his money, if he were to go by the routine DTPA soil test values. To elaborate, the soils from the *Ambalavayal* region have a Zn buffer power of 3.0358 as compared soils from the *Peruvannamuzhi* region which have a buffer power of only 0.7824. In other words, the *Amabalavayal* soils have the potential, which is early more than fourfold, to supply Zn as compared to *Peruvannamuzhi* soils. In practical terms, this would imply that the *Ambalavayal* soils would only require 25% of the total quantum of Zn applied to *Pervannamuzhi* soils. Translated into the recommendation, it would imply that the *Ambalavayal* soils would only need about 5 kg Zn fertilizer/ha when the *Peruvannamuzhi* soils require 23 kg/ha. In the case of *Thamarasseri* soils, the quantity required would be approximately 50% of the recommended rate. The pecuniary benefits to the poor black pepper farmer can easily be gauged from this sound and practical reasoning (Photos 5.2 and 5.3).

In fact, the scientists at the Indian Institute of Spices Research have made a gross underestimation of the Zn-supplying power of the soils in the State, because, their results and recommendations are all based on the soils of the *Peruvannamuzhi* region, where the black pepper research station is located. The soils here are acutely Zn impoverished, but, that is not the case in the entire State of Kerala. This also shows that, more often than not, practical recommendations from experiment stations, when extrapolated on a large-scale, can run into real problems, especially when such recommendations are based on routine soil testing procedures and the poor farmers could be at the receiving end of losing money. These results (Nair 2004) substantiate the earlier ones (Nair 2002) on the importance of precisely quantifying the Zn buffer power, rather than routinely estimate DTPA extractable Zn.

Photo 5.2 Black pepper in farmers' fields, Kerala State, India

Photo 5.3 Black pepper plantation in Kerala State, India

This approach is further discussed in the case of the author's investigations in Turkey (Central Anatolia), where severe deficiency is causing great adverse effects on what production. Farmers have been advised to apply as much as 100 kg $ZnSO_4$/ha, as a blanket recommendation, without obtaining any tangible and consistent response in wheat yield. Hence, the great economic importance of the "Buffer Power Concept," as against routine soil testing procedures, needs hardly any underscoring. Nair et al. (1997) have conclusively demonstrated the importance of the Buffer Power Concept in the case of potassium nutrition of cardamom (Elettaria cardamom M.), another very important spice crop, besides black pepper, in the State of Kerala (Photo 5.4).

5.3.2 *Central Asian Soils*

In a global study initiated by FAO and carried out in 30 countries, it was estimated that about 30% of the agricultural soils of the world are Zn-deficient (Sillanpaa 1982). Based on the analysis of 298 soil samples, Turkish soils were classified as Zn-deficient (Sillanpaa and Vlek 1985). Acute Zn deficiency triggering severe decreases in cereals yields have been noted in many countries of the tropical belt. Cereal species show large differential sensitivity to Zn deficiency. Large differences in sensitivity to Zn deficiency is also seen among genotypes within a specific cereal species (Graham and Rengel 1993). This indicates the existence of genetic potential to select and breed for Zn deficiency resistant genotypes. Among the cereals, wheat is particularly sensitive to Zn deficiency. Growing wheat on Zn-deficient soils leads

Photo 5.4 Shoot borer infestation in cardamom

to not only decrease in plant growth and yield, but would also lead to low Zn accumulation in grains. This will have a deleterious effect on human nutrition, because wheat grains are rich in phytate and fiber, which limit absorption and/or utilization of Zn in human cells (Welch 1993). Approximately 40% of the world's population suffer from micronutrients, including Zn, deficiency (Graham and Welch 1994). High consumption of cereal-based foods is thought to be one of the major reasons for Zn deficiency in humans. The proportion of wheat, maize, and rice consumed in diets of people in developing countries comprises about 90% of the average diet (Graham and Welch 1994). In Turkey, in Central Asia, it has been observed that Zn deficiency in humans is widespread where cereal-based foods dominate (Cavdar et al. 1983; Prasad 1982). Turkey is one of the major wheat producing countries in the world. Nearly 45% (4.5 million ha) of the wheat producing area in Turkey is located in the highly calcareous region of Central Anatolia. Central Anatolia is semi arid, being the driest region of Turkey. Scanty rainfall (about 325 mm) and low water supply greatly limit wheat yields. The author initiated a project to critically examine the role of the Buffer Power Concept in the Zn nutrition of wheat, as inconsistent results have been observed even with huge quantities (as much as 100 kg $ZnSO_4$/ha) of Zn application to the wheat crop.

Very recently, Nair (unpublished data) employed an adsorption– desorption equilibrium and a soil incubation technique to quantify the Zn buffer power of some highly Zn-deficient Turkish soils which have failed to show consistent response to Zn application based on DTPA extraction in wheat grown in these soils. Despite quite comparable DTPA extract- able Zn, the Zn buffer power of these soils varied largely from a low value of 27 to a very high value as high as 200. It is being hypothesized that the Zn buffer power could, possibly, hold the key to explaining the differential response of wheat to Zn application in these problematic soils. The technique developed by Nair (unpublished data) dispenses with the need to employ highly cost intensive EUF, which will certainly not be suitable to practical situations in many developing countries constrained by economic considerations, and in its place, employs a far simpler, and, simultaneously, far less costlier, technique of quantifying Zn intensity by an adsorption– desorption equilibrium technique using 0.01 M $CaCl_2$ solution or soil incubation technique.

Establishment of the Zn buffer power. The Zn buffer power curves are represented by the plot between the Zn "quantity" and Zn "intensity" parameters (Nair 1996). The "Q" and "I" parameters were established by laboratory procedures, one where an adsorption–desorption equilibrium technique was employed and another where an incubation technique was used, as described below.

Adsorption–desorption equilibrium technique to measure Zn "intensity". 5 g soil were weighed into PVC tubes saturated with 25 ml Zn solution ($ZnSO_4 \cdot 7H_2O$) in 0.01 M $CaCl_2$ solution in amounts ranging from 0 to 500 mg Zn/kg soil. The soil suspensions were shaken in an end-over-end shaker at ambient temperature (±25 °C) at 125 cycles/min. After 24 h the suspensions were filtered through "Blue Ribbon" 589 cube (3)— 150-mm filter paper (Schleicher and Schuell—German make). The soil on the filter paper was thoroughly washed with double distilled water and subsequently with 60% ethanol to remove any Zn ions not bound to the soil matrix and the entire mass of soil was transferred to another set of PVC containers. The filtrate was immediately analyzed for Zn using atomic absorption spectrophotometry.

Estimation of Zn "intensity". To the new set of PVC containers with the soil mass, 25 ml of 0.01 M $CaCl_2$ solution were added and they were shaken for 24 h under identical conditions, as explained above. After 24 h, the suspensions were filtered, as before, and the filtrate immediately analyzed for Zn. This represents Zn "intensity."

Estimation of Zn "quantity". The soil in the PVC containers was extracted with DTPA extractant, which is a mixture of 0.005 M DTPA, 0.01 M $CaCl_2$, and 0.1 M TEA, adjusted to a pH of 7.3 using 6 N HCl. The pH of the extractant was monitored using a portable pH meter. Twenty-five milliliters of the extractant was pipetted into the PVC containers and the soil suspensions were shaken as before, but, only for 2 h. The extraction period is crucial as originally suggested by the authors, since extraction is not complete after 2 h, the labile and nonlabile quantities of Zn continuing to dissolve. The filtrate was immediately analyzed for Zn by atomic absorption spectrophotometry and the Zn concentration in the filtrate was expressed as mg Zn/kg soil (ppm).

Each of the soil series had six Zn treatments, including a "No Zn" control treatment, and each treatment was replicated thrice to conform to a Completely Randomized Block Design. Thus, in all, in each cycle of operation, there were 54 samples and each was analyzed, at every stage, in duplicate.

The most important conclusions of great practical significance, emanating from these investigations, are the following. First, there is a very high correlation coefficient (0.8660***, significant at 0.1% confidence level) between Zn buffer power and percentage increase in wheat grain yield (Table 5.6). This unequivocally

Table 5.6 Interrelationship between Zn availability parameters and wheat grain yield and shoot tissue concentration of Zn (Turkey)

Parameter investigated	Correlation coefficient ("r" value)
DTPA soil test Zn versus Zn concentration in shoot tissue	0.672**
Zn buffer power versus percentage increase in wheat grain yield	0.860***

Notes: Note the very highly significant (at a confidence level of 99.9% (***)) "r" value between Zn buffer and percentage increase in wheat grain yield. Though DTPA soil test seems to have a positive influence on shoot Zn concentration, it is the Zn buffer power that yields far greater influence on wheat grain yield

Table 5.7 Effect of Zn fertilization (100 kg $ZnSO_4$/ha) on mean and range of Zn shoot tissue concentration and grain yield among 28 wheat genotypes grown in Konya soils and 25 grown in Gap soils in Turkey

	Tissue concentration (ppm)				Grain yield (kg/ha)				
	Without Zn		With Zn		Without Zn		With Zn		
Soil	Range	Mean	Range	Mean	Range	Mean	Range	Mean	Percentage increase
Konya	7–10	9	10–19	14	1550–4010	2710	248–5900	4570	69
Gap	16–23	19	22–31	25	1530–5260	3500	1550–5300	3510	0

Notes: The Zn buffer power is very clearly reflected in grain yield increase. Note that the "b" values (Zn buffer power) for both the adsorption–desorption equilibrium technique and incubation technique for Konya soils are 201 and 679, respectively. The corresponding values in *Koruklu* soils (Gap region) are 17 and 375, respectively. The percentage increase in wheat grain yield in Konya soils is 69 while that in *Koruklu* soils 0. In other words, the Konya soils are highly buffered for Zn availability

establishes that the single most important determinant in wheat yield increase is the soil's Zn buffer power. Both the experimental soils from Turkey were critically Zn-deficient (<1 mg Zn/kg soil, i.e., 1 ppm), while the German soil had 3.6 mg Zn/kg soil (3.6 ppm). And a uniform rate of 100 kg $ZnSO_4$/ha as $ZnSO_4{\cdot}7H_2O$, was applied to both soils. Yet, there was a huge margin of difference in the mean wheat grain yield (Table 5.7). The percentage increase in mean wheat grain yield in Konya soils, which are very highly Zn buffered, is 69 while those from Kuruklu, where the Zn power had a very low value, showed but negligent increase in wheat yield. The percentage increase coefficient of determination for the interrelationship between Zn buffer power and mean increase in wheat grain yield is 74. This simply illustrates that nearly 75% of yield variation can be explained by the Zn buffer power. The soil Zn dynamics which explains this phenomenon is as follows.

That Zn is principally transported through the diffusive mechanism is now an established fact and Nair (1984a, b) has provided experimental evidence where Zn buffer power is shown to be the most important criterion affecting the uptake process. Flux rates thus depend on Zn concentration gradients between root surfaces and the bulk soil solution (Nair 1992). Current results have to be viewed against this background. According to the model of Zn uptake proposed by Nair (1984a, b), the average concentration on the root surface is analogous to Zn "intensity" parameter, and so would critically affect the Zn uptake process. A root growing into the soil would, at first, encounter a relatively high Zn concentration at its surface, which would be more or less equal to the Zn concentration in the bulk soil solution. Due to root uptake of Zn, its concentration on the root surface would diminish. This decrease would not only depend on the uptake rate, but, also on the Zn buffer power, because, the Zn buffering mechanism would govern the rate of replenishment in the zone of Zn depletion around the roots. A soil with a high Zn buffer power would replenish the depleted soil much faster than a soil with low Zn buffer power. This differential mechanism is reflected in the Zn buffer power values (Table 5.3) of the soils from Konya and Kuruklu. Thus, it is logical to conclude that the uptake model proposed by Nair et al. (1984a, b) clearly explains the yield differential in the two soils investigated. The Konya soils, by virtue of their very high Zn buffer power, are able to maintain optimal Zn concentration on the wheat root surface and this positive impact is reflected in grain yield increase. The Kuruklu soils are unable to maintain the optimal Zn concentration on root surface, hence are unable to sustain high wheat yields (Table 5.7).

5.4 Case Studies with African Soils

Central African soils—the Republic of Cameroon. The Republic of Cameroon is a small country in Central Africa. Soils are lateritic, highly acidic, and soil pH ranges between 4.5 and 5.0. The country's agricultural scenario is confined mainly to cocoa, coffee, and rubber. Among the cereals, both rice and maize are popular. The country has a huge animal population, and good, and nutritious fodder is a great necessity of the country. With this view in background, an attempt was made to introduce white clover (*Trifolium repens* L.) from the United States, and the following investigations refer to the role of "The Nutrient Buffer Power Concept" in the nutrition of white clover.

Experimental soils. The soils investigated are widely varying. Their physico-chemical properties are listed in Table 5.8.

The main objective of this investigation was to examine how well "Buffer Power Concept" explains soil availability of P and K in very problematic African soils. And, white clover was chosen as the test crop because there is acute shortage of palatable and nutritious fodder for the livestock in the country. What is generally fed to the cattle is the post harvest waste of either rice or maize. This does not ensure good milk yield. It was thought that livestock industry had a tremendous potential in the country and white clover was chosen as the test crop. The seeds were obtained through a Memorandum of Understanding between The University Center, Dschang, The Republic of Cameroon, and the University of Florida (Institute of Food and Agricultural and Forest Sciences), in Gainesville, Florida. The investigation was confined to two soil types with very divergent physic-chemical characteristics (Table 5.8). Soil 2 is highly fertile compared to Soil 1.

Table 5.8 Physicochemical properties of the experimental soils

Details	Soil 1	Soil 2
Sand (%)	48	7
Silt (%)	18	41
Clay (%)	34	52
Organic carbon (%)	5.6	8.2
Total N (%)	0.37	0.48
pH (1 N KCl)	4.4	5.0
CEC (mequiv./100 g soil)	9.5	20.7
Base saturation (%)	21.8	32.2
Available P (ppm)	2.1[a]	13.0[a]
Available P (ppm)	3.7[b]	4.7[b]
Available P (ppm)	Tr.[c]	3.2[c]
Available K (mequiv./100 g soil)	0.18[d]	0.27[d]
Available K (mequiv./100 g soil)	0.17[c]	0.21[c]

Notes: All of the below-mentioned soil tests are "routine" soil tests for available P and K. The only exception is the 0.01 M $CaCl_2$ extraction which gives the P concentration in the equilibrium suspension of soil (Mattingly et al. 1963)

[a]Olsen's test

[b]Bray II test

[c]0.01 M $CaCl_2$ extraction

[d]NH_4OAc test

Table 5.9 Linear regression functions (Y = bx + a) and correlation coefficients ("r" values) for the interrelationship between Olsen's P test (Y = P "quantity") versus 0.01 M $CaCl_2$ extractable P (x = P "intensity"), and NH_4OAc extractable K (Y = K "quantity") versus 0.01 M $CaCl_2$ extractable K (x = K "intensity")

Nutrient	Soil	Regression function	Correlation coefficient ("r")
	1	0.41 x ± 0.24	0.994*
	2	0.47 x ± 2.37	0.890*
	1	0.17 x ± 0.10	0.960*
	2	1.28 x ± 0.17	0.940*

Notes: "b" in the regression function refers to the buffer power of the nutrient in question, a ¼ a constant

*Significant at a confidence level of 99.9%

Data in Table 5.9 clearly indicate that the experimental technique devised by this author has resulted in a remarkable positive interrelationship between the Q/I measurements. The very highly significant (at 99.9% confidence level) "r" values substantiates this contention. Also, soil 2 has a much higher P and K buffer power compared to soil 1. What follows is the further examination of the question whether integrating the P and K buffer power into the computations enhances predictability of the P and K concentration in plant top, and, eventually, P and K uptake. Answers to this question are presented in Table 5.10.

Table 5.10 Correlation coefficients for plant top concentration and plant top uptake of P and K without (A) and with (B) the P and K buffer of the experimental soils

Details	A	B
P concentration versus Olsen's soil test	0.61*	0.69*
K concentration versus NH_4OAc soil test	0.05	0.62*
P uptake versus Olsen's soil test	0.17	0.77*
K uptake versus NH_4OAc soil test	0.18	0.78*

*Significant at a confidence level of 99.9%

Data in Table 5.10 unequivocally establish the important fact that integration of the buffer power, both in the case of P and K, enhances predictability of P concentration in plant top and P uptake, so also K concentration in plant top and K uptake, in a very highly significant (at a confidence level of 99.9%) manner. More than this, there are other very important additional features of the results, which follow. That is, first, inasmuch as the P and K uptake process is concerned, neither of the widely practiced routine soil test for P, that is the Olsen's soil test, nor the NH_4OAc test for K, showed any positive statistically significant relationship. This is reflected in the very low "r" values (Table 5.10). But, most remarkably, these "r" values shoot up to very highly significant ones (at a confidence level of 99.9%) when the corresponding P and K buffer power values are incorporated into the computations. In fact, in the case of Olsen's soil test, for P uptake, the coefficient of determination increases from a mere 2.9% in the absence of the P buffer power to 59.3% when the corresponding P buffer power value is also integrated into the computations. In the case of the routine NH_4OAc extraction for K uptake, the corresponding values are 3.2% and 60.8%. The coefficient of determination is a precise measure of the accuracy of prediction. In other words, these results unequivocally demonstrate that by integrating the respective P and K buffer power values into the computations, a substantial variance in P and K uptake by plant top can be explained. This is, indeed, a very remarkable observation of this investigation, which has considerable significance in white clover production in Cameroon. This is, because, once the P and K buffer power values of the soils in question are determined, to begin with, the application of the important plant nutrients through their respective carriers (chemical fertilizers) could be appropriately designed. This has very extensive practical consequence in white clover production in Cameroon. Most remarkably, in the case of K, the coefficient of determination for plant top concentration increased from a mere 0.25% to 38.4%. This translates into a 15,360% increase in variance prediction.

Results in Tables 5.11 and 5.12 have an important bearing on the practical aspects of white clover production in Cameroon, *vis-à-vis* "The Nutrient Buffer Power Concept." The most important conclusion is that among the two nutrients investigated, namely, P and K, response to P fertilizer application has shown a consistent pattern, with both higher dry matter production and P uptake. With enhanced P fertilizer application rates, there has been consistent increase in dry matter production and P uptake. The magnitude of dry matter production and P uptake in response to enhanced P fertilizer application has been more in the case of soil 2 as compared to soil 1. When we come to K fertilizer application and crop response, it has been

Table 5.11 Plant top dry matter production (kg/ha) and P uptake (kg/ha) from the experimental soils as a function of P fertilizer application

	Dry matter production (kg/ha)				P uptake (kg/ha)			
	P application rate (kg/ha)				P application rate (kg/ha)			
Soil	0	12.5	25	37.5	0	12.5	25	37.5
1	29	62	78	76	43	95	152	191
2	67	96	99	93	106	154	176	181

Table 5.12 Plant top dry matter production (kg/ha) and K uptake (kg/ha) from the experimental soils as a function of K fertilizer application

	Dry matter production (kg/ha)				K uptake (kg/ha)			
	K application rate (kg/ha)				K application rate (kg/ha)			
Soil	0	50	100	200	0	50	100	200
1	32	24	27	20	1371	959	1071	737
2	65	73	76	87	2989	3675	3815	4576

quite inconsistent. The consistent and better response in soil 2 as compared to soil 1 can be traced to its much better buffer power (Table 5.9).

What are the implications of this finding in white clover production in Cameroon?

The Republic of Cameroon is an emerging and stable country (bilingual—both French and English) on the African continent, with a small population, but, with vast natural resources like oil, minerals (especially bauxite, raw material for aluminum, which is much sought after by aerospace industry), agriculture (specially cocoa production, second highest producer on the African continent next to Ghana), and a vast potential for tourism. It has a stable livestock industry (cattle production) which has a very bright future. Though on the countryside of the natural grazing is the common route for cattle feeding, things are giving way to supplemental fodder feeding. It is here the potential for white clover cultivation emerges. This author imported the seeds of white clover from the United States and the results of the experiments reported here are from the imported seeds. The objective behind this investigation was to popularize the cultivation of white clover in Cameroon and it is seen that the "Buffer Power Concept" has a major role to play in its nutrition, inasmuch as P and K nutrition of the plant is concerned, as could be clearly deduced from the experimental results discussed above.

This research formed part of a Master's thesis of one the author's students, Ms. Che Felicitas, while he held the position of Professor and Head of the Department of Soil Sciences at The University Center, and was selected as the best thesis submitted to the University for the Master's degree and she received a sum of 400,000 CFA (equivalent to more than U.S. $1000) awarded in recognition of the academic excellence of the research by the organization for advancement of science on the African continent. The award was given by the Minister of Higher Education and

Scientific Research, on behalf of the President Paul Biya, of The Republic of Cameroon, and was celebrated as a very important national event.

References

Barber SA (1984) Soil nutrient bioavailability: a mechanistic approach. Wiley, New York

Cavdar AO, Arcasoy A, Cin S, Babacan E, Gozdasoglu S (1983) Geophagia in Turkey: iron and zinc deficiency, iron and zinc absorption studies and response to treatment with zinc in geophagia cases. In: Zinc deficiency in human subjects. Alan R. Liss Inc, New York, pp 71–79

Drew MC, Nye PH, Vaidyanathan LV (1969) The supply of nutrient ions by diffusion to plant roots in soil. I. Absorption of potassium by cylindrical roots of onion and leek. Plant Soil 30:252–270

Elgawhary SM, Lindsay WL, Kemper WD (1970) Effect of EDTA on the self-diffusion of zinc in aqueous solution and in soil. Soil Sci Soc Am Proc 34:66–70

Graham RD, Rengel Z (1993) Genotypic variation in zinc uptake and utilization by plants. In: Robson AD (ed) Zinc in soils and plants. Kluwer Academic Publishers, Dordrecht, pp 107–118

Graham RD, Welch RM (1994) Breeding for staple—food crops with high micronutrient density: long-term sustainable agricultural solutions to hidden hunger in developing countries. In: IFPRI workshop on food policy and agricultural technology to improve diet quality and nutrition, Annapolis, Maryland, January 10–12

Marschner H (1994) Micronutr News Inform 14:3–5

Mattingly CEG, Russell RD, Jephcott BM (1963) Experiments on cumulative dress-ings of fertilizers on calcareous soils in Southwest England. II Phosphorus uptake by rye- grass in the green house. J Sci Food Agric 14:629–637

Nair KPP (1984a) Towards a better approach to soil testing based on the buffer power concept. In: Proc. 6th int. colloquium for the optimization of plant nutrition, 2–8

Nair KPP (1984b) Zinc buffer power as an important criterion for a dependable assessment of plant uptake. Plant Soil 81:209–215

Nair KPP (1989) Comments on "phosphorus supply characteristics of 33 soils as influenced by seven rates of phosphorus addition" soil Sci. Soc Am J 53:984

Nair KPP (1990) Quantification of P and Zn replenishment capacity of soils as a dependable alternative to routine soil tests in assessing their bio-availability to growing plants. In: Transactions of the 14th International Congress of Soil Science, Kyoto, Japan, August 12–18, Comm. IV, pp 493–494

Nair KPP (1992) Measuring P buffer power to improve routine soil testning for phosphate. Eur J Agron 1(2):79–84

Nair KPP (1996) The buffering power of plant nutrients and effects on availability. Adv Agron 57:237–287

Nair KPP (2002) "The nutrient buffer power concept" – a revolutionary soil management technique for sustainable agriculture. Invitational research paper presented at the 17th world soil science congress held in Bangkok in August 2002

Nair KPP (2004) Sustaining crop production in the developing world through "the nutrient buffer power concept" – a case study with black pepper (Piper nigrum) growing at low pH in the laterite soils of India proc. of the 16th international symposium on plant- soil interactions at low pH, Sendai, Japan, August 1–5, 392

Nair KPP, Babu GR (1975) Zinc–phosphorus–iron interaction studies in maize. Plant Soil 42:517–536

Nair KPP, Prabhat G (1977) Differential response of tropical maize genotypes to zinc and manganese nutrition. Plant Soil 47:149–159

Nair KPP, Ram N, Sharma PK (1984) Quantitative relationship between zinc transport and plant uptake. Plant Soil 81:217–220

Nair KPP, Ram N, Sharma PK (1984a) Quantitative relationship between zinc transport and plant uptake. Plant Soil 81:217–220

Nair KPP, Ram N, Sharma PK (1984b) Quantitative relationship between zinc transport and plant uptake. Plant Soil 81:217–220

Nair KPP, Sadanandan AK, Hamza S, Abraham J (1997) The importance of potassium buffer power in the growth and yield of cardamom. J Plant Nutr 20(7 & 8):987–997

Nemeth K (1979) The availability of nutrients in the soil as determined by elec-troultrafiltration (EUF). Adv Agron 31:155–188

Nemeth K, Recke H (1982) Reproducibility of Ca, K, Na, and P contents in the different EUF fractions. Plant Soil 64:25–33

Prasad AS (1982) History of zinc in human nutrition. In: Prasad AS, Dreosti IE, Hetzel BS (eds) Clinical applications of recent advances in zinc metabolism. Alan R. Liss Inc., New York, pp 1–17

Shuman LM (1975) The effect of soil properties on zinc adsorption by soils. Soil Sci Soc Am Proc 39:454–458

Sillanpaa M (1982) Micronutrients and the nutrient status of soils: a global study, FAO soils bulletin no. 48. FAO, Rome

Sillanpaa M, Vlek PLG (1985) Micronutrients and the agroecology of tropical and mediterranean regions. Fertil Res 7:151–167

Welch RM (1993) Zinc concentrations and forms in plants for humans and animals. In: Robson AD (ed) Zinc in soils and plants. Kluwer Academic Publishers, Dordrecht, pp 183–195

Wilkinson HF, Loneragan JF, Quirk JP (1968) The movement of zinc to plant roots. Soil Sci Soc Am Proc 32:831–833

Chapter 6
Concluding Comments and Future Imperatives

Abstract Historically, soil testing has been used to quantify bio availability of essential plant nutrients to field grown crops. However, contemporary soil tests are based on philosophies and procedures developed several decades ago without significant changes in their general approach. For a soil test to be accurate, one needs to clearly understand the physico-chemico-physiological processes at the soil-root interface, and, an understanding of soils and plant root systems as polycationic systems is essential. It is this knowledge that leads to sound prescriptive soil management practices inasmuch as nutrient bioavailability *vis-à-vis* fertilizer application is concerned. The book presents a revolutionary new idea, now globally known as "The Nutrient Buffer Power Concept", based on the thermodynamics of soil nutrient bioavailability that opens up a new chapter in the domain of soil testing. The success of a new approach, to a large measure, rests with the ingenuity of those applying it to suit the demands of a new situation. "The Nutrient Buffer Power Concept" project was shortlisted for the very prestigious US $1 Million 2012 Rolex Awards For Enterprise for its originality from more than 3500 nominations worldwide and is the only project selected for this coveted distinction from the Asian continent.

Historically, soil testing has been used to quantify availability of essential plant nutrients to field-grown crops. However, contemporary soil tests are based on philosophies and procedures developed several decades ago without significant changes in their general approach. For a soil test to be accurate, one needs to clearly understand the physico–chemico–physiological processes at the soil–root interface, and an understanding of soils and plant root systems as polycationic systems is essential. It is this knowledge that leads to sound prescriptive soil management practices inasmuch as nutrient availability vis-a-vis fertilizer application is concerned, because of all the factors which govern sustainability of crop production, the nutrient factor is the most important, yet, it is also the least resilient to management. This review focuses on the buffering power of plant nutrients, especially phosphorus, potassium, zinc, and some other heavy metals such as cadmium, on their availability to growing plants. Experimental evidence is presented from research carried on temperate (Central European and North European) and tropical soils (African and

© The Author(s), under exclusive license to Springer Nature Switzerland AG 2021
K. P. Nair, *Thermodynamics of Soil Nutrient Bioavailability*,
https://doi.org/10.1007/978-3-030-76817-1_6

Asian soils) which unequivocally establishes the fact that precise quantification of the nutrient buffer power holds the key to a clearer understanding of the plant availability of some of the more important essential plant nutrients, such as phosphorus, potassium, and zinc. The Nutrient Buffer Power Concept attempts to explain availability of plant nutrients on the basis of the diffusion model, as it is the principal process by which phosphorus, potassium, and zinc are taken up by plant roots. Possibly, other nutrients which are taken up by plant roots by the same process would also conform to the principles of this concept. To substantiate this, supportive experimental evidence is needed.

There are similarities between the buffer power approach and quantity– intensity (Q/I) approach in predicting plant availability of important nutrients such as phosphorus and potassium. While the Q/I approach has met with a fair measure of success in predicting P availability, with regard to K availability the Q/I relationship does not enjoy universal acceptance since a single relationship for all soils between K uptake by a given crop and AR_K does not exist, perhaps due to the nature of the soil components regulating AR_K.

In substance, soil testing in a laboratory is meant to simulate what a plant root would accomplish in a field soil. All soil tests meet this requirement with different degrees of success. But, for a soil test to be universal in scope, it must attempt to quantify, as precisely as possible, factors which most crucially affect plant root uptake. Of all the factors involved in nutrient uptake, the most crucial is the average nutrient concentration on the root surface. In a dynamic state of plant growth it is virtually impossible to quantify precisely the average nutrient concentration on the root surface. The nutrient's buffer power is an indirect measure of the average nutrient concentration on the root surface, and analytical methods which quantify the buffer power with precision would render the soil test in question dependable; hence, the fertilizer recommendation which follows would be reproducible. It is against this background that The Nutrient Buffer Power Concept merits wider application for field use. Fertilizer prices have dramatically escalated following the oil crisis of the mid-1970s, and field experience shows that recommendations emanating from routine soil tests have met with only limited success. A number of countries on the Asian and African continents spend enormous sums of money on importing fertilizers, which is a great strain on their national exchequers. Most of these countries have no infrastructure for fertilizer manufacture and where there is any, the soil testing support service is far from satisfactory. Since fertilizer is a crucial input, second only to water, in modern farming, cost-effective fertilizer recommendations are an important prerequisite for successful agriculture. In countries such as India, fertilizer prices, especially of phosphate and potassium, have skyrocketed following governmental policies centered on a market economy. In such situations, farmer confidence can only be instilled and maintained by accurate soil testing and reproducible fertilizer recommendations.

The success of a new approach, to a large measure, rests with the ingenuity of those applying it to suit the demands of a new situation. This principle is no exception in making the "Nutrient Buffer Power Concept" a universal success. As compared to routine soil testing, this approach calls for the accurate determination of the buffer power of the nutrient in question, to begin with. Once this is done, the buffer power factor can be integrated with the routine soil test data and fertilizer recommendations can be made on the basis of this new information. This implies that, in addition to obtaining routine soil test data, one also needs to know the buffer power. This review has shown how the buffer power is determined based on analytical methods varying from a simple adsorption–desorption equilibrium technique to elaborate EUF. Only selected laboratories around the world have EUF facilities. Where they are absent, alternative analytical procedures to precisely quantify the buffer power need to be developed. Developing appropriate procedures to quantify the buffer power and modification of the protocol for routine soil testing to integrate buffer power data for precise fertilizer recommendations calls for both ingenuity and drive on the part of the researcher, but that will be the future imperative of accountable soil testing.

6.1 Input Substitution and System Redesign

An important point that emanates from the above narrative is this: Chemically centric—highly chemical fertilizer dependent and highly soil extractive farming, euphemistically known as the "green revolution" is no more sustainable, an experience of the last half a century, globally demonstrated. Equally so, the call for a "totally organic farming" mode will not result in adequate food production for a burgeoning global population. If we have to move forward one has to find a mid path where the source of crop nutrition will have to be balanced by a healthy combination of both the chemical inputs and adequate organic inputs. Alternatively, the approach has to zero on precise nutrient requirement for the major crops as the foregoing narrative has very clearly demonstrated. However, a discussion on the identification of the main conceptual paradigms of organic farming is central to, social, agronomic, soil scientists, including those engaged in fertilizer management and field extension work. Based on the research done in Denmark, Kaltoft (1999) identified four paradigms: the paradigm of nutrients (or conventional point of view), the paradigm of soil fertility (ecological point of view), the dynamic paradigm of communication which involves aspects of the three others. In the case of the Netherlands, other authors distinguish three conceptions within organic farming: natural as the organic, natural as the ecological and natural as referring to nature as a holistic entity (Verhoog et al. 2003). They relate these conceptions to three main approaches to organic agriculture: a no chemicals approach, an agro-ecological approach and an integrity approach. This last one entails a spiritual dimension, neighboring the holistic biodynamic principles and, in a way, conforms to the idea of soil as the "**soul of infinite life**" (Nair 2019a, b). A third study analyses the organic farmers' relations to nature and places them between an anthropocentric pole and a "natural pragmatic" pole (Tress 2001). In the first case, productivity is the guideline for farmers' actions and nature has only an instrumental value; farmers aim for a direct control of agricultural production and performances in an unchanged technological frame of reference. In the second case, humans and nature co-exist, nature is recognized as having a certain value and as being an autonomous entity, interactions between techniques and the components of the "agroecosystem" have to be built so as to enhance natural regulation processes and partial or indirect effects.

All of the above suggestions of distinct categories of organic conceptions can be related to the binary distinction between an "input substitution paradigm" and a "system redesign" paradigm first suggested by biologists (Hill 1985; Hill and MacRae 1995). The first defines organic farming as the ban of certain inputs and/or the recommendation of the others (list of eligible inputs to "fight against" pests and diseases, although with biological methods), whereas the second one defines it through more qualitative and global principles, such as, crop rotation or means to overcome sanitary problems. This holistic paradigm refers more broadly to the construction of diversified production systems following the ecological model considered as the "natural" one, where interactions between components guarantee fertility, productivity and resilience properties.

The transition of a farm inside organic agriculture, once converted, could be defined through the paradigms considered as successive steps. Indeed, by definition, organic agriculture relies on input substitution. Once this substitution is accomplished, it is possible to aim for a better efficiency of inputs inside the organic framework, or, to go even further and to aim for a "redesign".

Does the notion of redesign also reflect the parallel evolution of farm and landscape structures/organic farms are supposed to be diversified and this diversification is to be seen in landscape changes. Conversion can then be considered as a case study to test land use options and crop combinations.

Many Danish investigations shed light on the above-mentioned aspect. Langer (2002) used a database composed of 448 conversion projects in which farmers describe their situations and their projects regarding crops and livestock production. Over a four-year period, no tangible evolution was perceived in production orientations. The regional distribution of cropping systems would therefore not be strongly modified in the short term. Conversely, land-use patterns change with conversion: a 20% decrease in the acreage for cereal production, a doubling of the acreage for grassland in rotation, a decrease in acreages for set-aside, row crops and oilseed rape. However, these farmers' intentions have not been verified through monitoring after conversion.

From the analysis of comprehensive interviews with 150 farmers, the geographer Guthman (2000) used methods based on indicators elaborated by biotechnical sciences (Van der Werf and Petit 2002), so as to classify the farmers according to their degree of adoption of agro-ecological principles (Altieri 1995), and to the dynamics of conventionalization. The analysis, which looks both at the differences between organic farmers, and, at the gaps between agro-ecological principles and farmers' practices, leads to a double distinction as follows:

1. Between mixed farmers who partially converted to organic farming and restrict themselves to following the rules which they interpret in terms of input substitution, and, farmers who fully converted
2. Between large farms and smaller farms, the large farms depend more on external inputs, whereas the small ones rely on compost-making and intercropping

However, only few of the producers approximate an agro-ecological ideal integrating crops and livestock, companion crops, and/or an intense mosaic of cropping design combined with a high degree of on-farm-input development. As for the pest and disease management, the methods used range from proactive prophylactic measures and habitat maintenance enhancing beneficial insects to an organic agriculture "by default" without an explicit strategy to manage or by-pass biotechnical issues (apart from using eligible inputs). In between these polarities, a "wait and see" behavior can be identified, quite similar to an IPM (Integrated Pest Management) strategy.

The distinction between the two paradigms of input substitution and system redesign converges with the distinction identified in the conventionalization debate between two distinct trends in organics, opposing industrial farmers and farmers involved in a more ideological and agro-ecological vision of organic farming. The concept of "conventionalization" was developed by political food economists in a founding article in 1997 (Buck et al. 1997). According to them, organic production and markets were being overcome by large-scale structures and were becoming more and more vertically integrated, just like in conventional industrial type agriculture. Conventionalization is characterized by the concentration of capital among fewer and larger growers and intermediaries more equipped to deal with retailers, the erosion of organic standards, the generalization of the substitution of allowable inputs for proscribed inputs, and, a growing dependence of farmers on input suppliers and supermarkets. This evolution led to a bifurcation into two distinct trends: a main conventionalized one and a minor resistant one.

The conventionalization thesis was developed and is still debated over California, where large-scale growers and monoculture are prevalent. In the United Kingdom (former, now Britain) the growth of the organic sector has also generated a decline in farm gate prices and farmers' income which has called into question the sustainability of current levels of organic production (Smith and Marsden 2004). This phenomenon is due to the conjunction of private-sector initiatives (consumption-oriented) and government incentives to increase conversion (production-oriented), which lead the authors to highlight the limits of government intervention when it only deals with production and neglects the downstream supply chains. Studies in Australia (Lockie and Halpin 2005), Canada (Hall and Mogyorody 2001) and southern countries (Halberg 2007) show that more diverse evolutions should be acknowledged. If the conventionalization thesis reveals how the boundaries between conventional and organic agriculture might be contested and mobile (Goodman 2000), it is therefore commonly agreed that the analyses have to be clearly related to their specific geographical context.

6.2 Organic Farming as an Indicator of Society Questioning Current Agriculture and Food Models

A very important aspect of current global agriculture, in the context of vast degradation of soil resources, globally, in particular in South Asia, specifically India, due to soil nutrient usage mismanagement (Nair 2019a) is how the society is questioning the current conventional agricultural systems. If social sciences take an increasing part in this debate around the paradigms sustaining the conceptions and practices of organic farming, discussed above, their specific contribution also involves the impact of the "organic movement" on civil society and the links to a broader debate over the future of agriculture and food systems.

Undoubtedly, the role of the organic movement to day is so vastly different than what it was in the beginning. Possibly, the most important reason for this is the huge environmental hazards the highly soil exploitative farming, euphemistically called the "green revolution", which caused to both the soils (huge soil degradation, Nair 2019a) and the environment (enhanced global warming, Nair 2019b). Earlier, the focus was enhanced food production, ironically at any cost, as had happened in India, in the State of Punjab, the "cradle of green revolution". Additionally, the farmers' greed to harvest more and more profit was also a big driving factor. In fact, Punjab not only lost thousands of acres of fertile land due to soil degradation, where it became very evident that not even a blade of grass will grow without huge investments in soil reclamation, but, also, became the hotbed of cancer due to unbridled use of herbicides, insecticides and fungicides, to prop up both wheat and rice yields. The Gurdaspur district in Punjab State, where this unbridled use of herbicides, insecticides and fungicides continued for decades became known as the "Capital of Cancer". Indian experience is still rooted in the "fertilizer centric" farming, though, organic farming is catching on because of the widespread environmental hazards due to the green revolution in different states. For instance, in the State of Kerala, in southern India, the official agricultural policy is organic farming, though practiced only in patches in different districts. There is, as of now, no other State in India, except Kerala State, where organic farming is a state official policy. The principal concern of the farmers is that yields are much lower, and, it is not often that they can sell their produce at a premium since markets are not well organized.

When we take a global view and find that the role of organic movement is surely different today than in its beginnings, when the mission of agriculture was to feed the nations and when productivism was, therefore, legitimate, with hardly any place for social criticism and movements in agricultural decisions (MacMahon 2005). In the French case, it sees that the organic pioneers failed to create a real social movement which would have brought together producers and consumers (Piriou 2002). According to this author, farmers followed the logic of their corporate lobby and could not go beyond the boundaries of their profession. Almost two decades ago, however, from the analysis of the investigations of organic farming carried out between the seventies and eighties, Barres et al. (1985) identified a progressive shift towards a willingness to be accepted and approved by society and to develop links between cities and the countryside. However, it is only very recently that an alliance between producers and consumers really began to visibly take shape, even though the importance of pioneering signs and experiences such as organic cooperatives networks has to be recognized.

Presently, organic networks greatly contribute to getting the civil society involved in debates over agriculture and food. The spate of conferences, workshops etc., at national and international level is testimony to this important fact. In fact, if one takes the example of the State of Kerala, in, India, the official policy of the State Government is organic farming, though, it must be admitted that many farmers, big and small, in the State, practice, clandestinely heavy chemical fertilizer-based farming, mainly targeting the yield factor. Their argument is "to produce more" and sell it relatively cheap unlike the ill organized organic farmers who produce less and demand a premium on their produce, which they claim as "chemical free". And, these farmers are not well organized in cooperatives or other forms, unlike in Europe. But, what helps them is an increasingly "chemically conscious" clientele, who would, at least the rich among them, are willing to pay a premium for the "chemically free" organic produce. This population segment is very health conscious. But, the vast majority of the Keralites would rather pay less and ready to take a risk consuming food enriched by inorganic chemicals. But, for sure, a strong sentiment in support of organic farming has been built with ample support from the State Government.

In Europe, certain authors denounce a repression of the organic movement's ideological content and a betrayal of organics' initial ideals through which the dominating definition of organic agriculture reduces relations to nature to a question of "allowed inputs". Both the institutionalization and conventionalization of organics go hand in hand with a decline in its idealistic vision and decline in farmers' environmental concerns (Kaltoft 1999). Many authors, though, have a more optimistic vision of the possible place of organics in political and social change. This place would rely on the capacity of organics to transform the way people look at their food, in particular, through unveiling the conditions of production, instead of hiding them. Alternative networks invite consumers to consider products through the way they are produced, the kind of labor involved relationship to nature, or even public research investments, which are implied in this production, and, not only through the material characteristics of products. Thus, organics could be an answer, among others, to larger concerns about the rising centralized control of society's relations with agricultural nature (Goodman 2000).

6.3 The Study of Transitions in Agriculture: Beyond Disciplinary Divisions

Presently, it is becoming extremely clear that it is necessary to explore and develop intermediary forms between conventional agriculture, whose thrust simply on productivity and profitability proves to be simply unsustainable, socially, and environmentally hazardous, as has been demonstrated by the highly fertilizer-centric farming in India, in particular, and South Asa, in general, euphemistically known as green revolution with its highly deleterious environmental fallout—soil degradation, global warming, displacement of native cultivars etc. (Nair 2019a, b) and an organic agriculture which is not suitable for every soil, region, and the farmer. In this context, the two paradigms of input substitution and system redesign, that were identified on the basis of organic investigations, might help enrich and frame the discussion over future agricultural models. Beyond organic farming, many investigations use these paradigms to analyze the transformations in agriculture.

In the Brazilian State of Rio Grande do Sul, these paradigms help to classify all farmers into three categories: (1) conventional farmers (who reduce their use of chemical input for economic reasons) (2) farmers in transition (who substitute chemical inputs with biological ones) and (3) farmers in a "redesign" category of transition (who adopt an ecological and systemic approach incorporating soil ecological management, crop rotation and diversification, mixed crop-livestock integration, reforestation, and management of agro-forestry systems. According to certain authors, integrated production also corresponds to a mid-term and fluid transition between conventional and organic agriculture (Niggli 1999), with strategies such as integrated pest management (IPM), reasonable crop fertilization and prioritization of biological control methods (Hodges 1981). Such categories can also be found in India to varying degrees (Nair 2019a).

Taking into account the complexity of transitions in agriculture supposes going beyond a restricted agronomical vision which would ignore the social aspects linked to food systems as well as the natural phenomena such as climate change. The latter aspect has been elaborately explained in the book "Combating Global Warming—The Role of Crop Wild Relatives For Food Security", Springer (Nair 2019b). This is what some authors try to do by defining agro-ecology broadly as an ecology of food systems (Francis et al. 2003) based on the notions of ecosystems and agroecosystem (Gliessman 1990). and, which can be opposed to a vision of agro ecology as an ecology applied to agricultural production (Altieri 1983). Interdisciplinarity is considered by these authors as a prerequisite to the development of agroecology, as economic and social aspects have to be to be included (Dalgaard et al. 2003). In this regard, the role of sustainable soil management based on a revolutionary fertilizer management technique now globally known as "The Nutrient Buffer Power Concept" and elaborately explained in the book "Intelligent Soil Management For Sustainable Agriculture—The Nutrient Buffer Power Concept", Springer (Nair, 2019a) assumes much relevance.

Interdisciplinary approaches are more generally of great interest and relevance when considering conversion to organic farming as an exemplary case of larger transitions in agriculture, as they allow the taking into consideration of these transitions by overcoming their sole technical aspects, and, by paying greater attention to their temporal dimensions. With this perspective in mind, approaching conversion to organic farming from the crossroads of several different disciplines holds promise.

To cross history with agronomy would allow an analysis over time of the transformation of the historical paradigms of agronomy, which can be identified as successively chemical, physical, biological, and, finally ecological paradigms (Robin and Aeschlimann 2007). Specifically for organic agriculture, a historical approach can help to identify the foundation paradigms and their differences, mainly, between Steiner's organic vision of a farm integrating breeding as a key component, and Howard's vision favoring soil fertility and humus management. This leads to renouncing the idea of a unique original paradigm for organic farming and contributes to the necessary acknowledgement of organics' internal diversity (Besson 2007).

The collaboration between agronomy of production and sociology allows the study of the transformations of farmers' practices over time. The notion of a trajectory allows the consideration of conversion over a longer period of time and from a wider point of view than in its official definition. Organic farmer' trajectories encompass at the same time technical trajectories, social trajectories, learning trajectories and network trajectories. Compared with the classical typological methods, such studies, involving agronomists and social scientists, prove that the interdisciplinary approaches are more adequate to reflect the actual diversity of organic farming, as well as the evolutionary potential of farming systems and transitional nature of converging trajectories. In these socio-technical trajectories, the questions of plant protection practices, of input use and farmers' representations of these issues, are central. It may seem paradoxical to note that these questions are not often raised in investigations of organic farming, may be because these problems are supposed to be solved, whereas they are more often central to comparative studies of conventional and organic farming. Pesticides have been proved to be not only a crop protection means, but also, a way to maintain a visible standard, in the sense that conventional farmers are often proud to have clean fields, which is acknowledged as a major bottleneck regarding the adoption of low-input agricultural practices.

Investigation in organic farming as an innovation is of prime interest when thinking in general terms about the transition towards more sustainable agriculture. Some investigators have applied to the case of organic farming the classical theories of adoption and diffusion, which identify innovators, as well as early and late adopters throughout the diffusion of any innovation (Padel 2001). As organic farmers do not always come from an agricultural background and are often lined to scientists and/or environmental and consumer movements, other types of innovation than classical top down processes can be expected (Michelsen 2001). The construction of knowledge and the co-production of innovations is still quite an unexplored area of research (Gibbons et al. 1994), even though there are pioneering works which it might be useful to go back to (Salmona 1994).

In sum, it must be sad that, the interactions between agricultural sciences and social sciences are a key to enlarge the scope from agriculture to food chains, and, thereby, consider the sustainability of organic farming or any other form of farming. Other investigators, as well, suggest consideration of the different development models of organic farming. Sylvander et al. propose to identify these developmental models through two axes, the first one opposing individual logics to collective organization forms linked to markets and territories, the second one opposing the mere compliance with organic farming rules to a re-conception of systems.

6.4 The Concept of Soil Health

Lehmann et al. (2019) has in an online publication discussed the concept of soil health, as follows.

> Soil health is the continued capacity of soil to function as a vital living ecosystem that sustains plants, animals and humans, and connects agricultural and soil science to policy, stakeholder needs and sustainable supply-chain management. Historically, soil assessments focused on crop production, but, today, soil health also includes the role of soil in water quality, climate change and human health. However, quantifying soil health is still dominated by chemical indicators, despite growing appreciation of the importance of soil biodiversity, owing to limited functional knowledge and lack of effective methods. In this Perspective, the definition and history of soil health are described and compared with other soil concepts. We outline ecosystem services provided by soils, the indicators used to measure soil functionality and their integration into informative soil-health indices. Scientists should embrace soil health as an overarching principle that contributes to sustainability goals, rather than only a property to measure.

Soil is a complex system at the intersection of the atmosphere, lithosphere, hydrosphere and biosphere that is critical to food production and key to sustainability through its support of important societal and ecosystem services. It is in this context that the concept of soil health emerged in the early 2000s and, today, has linkages to the emerging 'One Health' concept, in which the health of humans, animals and the environment are all connected.

The terminology, concept and operationalization of soil health are still evolving (Box 1). It is now defined by most agencies, such as the US Department of Agriculture, as "the continued capacity of soil to function as a vital living ecosystem that sustains plants, animals, and humans" (https://www.nrcs.usda.gov/wps/portal/nrcs/main/soils/health/). Several other related concepts exist, including soil fertility, soil quality and soil security (Fig. 6.1), which also emphasize the role or functioning of soil in society, ecosystems and/or agriculture.

The narrowest of these terms is soil fertility, which refers to the role of soil in crop production. Soil fertility is managed by farmers at the field scale for the purpose of cost effective crop production and entirely focuses on growing food, fuel and fiber for human use.

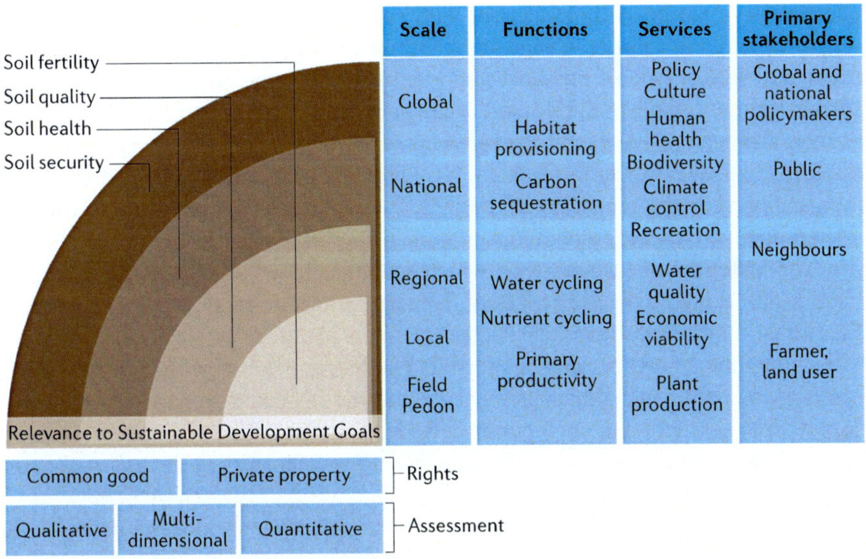

Fig. 6.1 Soil fertility, quality, health and security. The concepts vary by what relevant spatial scales, functions, ecosystem services and stakeholders they capture (listed as nested concepts on the right of the figure). The concepts also differ in the view of soil rights and assessments. Soil health encompasses a broad range of ecosystem functions, services and actors, impacting a wide array of sustainability goals. The five functions listed here impact overall soil-ecosystem services

Soil quality is the historic origin of the term soil health and describes the ability of a soil to function for agriculture and its immediate environmental context. Soil quality, therefore, includes soil effects on water quality, plant and animal health within entire ecosystems. Although the terms are often used synonymously, we argue that soil health is distinct from soil quality, as the scope of soil health extends beyond human health to broader sustainability goals that include planetary health, whereas the scope of soil quality usually focuses on ecosystem services with reference to humans.

Soil security, introduced in 2012, is the most recent and broadest term of the four, and encompasses soil health, using the term 'soil condition' to describe the manageable properties of soil. Soil security relates to the need for access to soilecosystem services to be on the same level as other human rights, and is, therefore, often used in a policy context, encompassing human culture, capital and legal aspects of soil management. Importantly, soil security allows for productive conversation about soil as a common good, similar to water and air, rather than only as private property (as in soil fertility and quality). We believe that this view must be moved to the center of the debate about the role of soils in sustainability and governance.

Soil health encompasses scales, stakeholders, functions and assessment tools relevant to soil quality and fertility, and shares some of the policy dimension of soil security (Fig. 6.1), going beyond a focus on only crop production or other explicitly human benefits. The multidimensionality of the soilhealth concept allows for soilmanagement goals to be aligned with sustainability goals, and should provide the foundation to consider a large number of stakeholders, functions, and spatial and temporal scales. One of the most important achievements of the soilhealth framework (initially under the term soil quality) is the addition of an urgently needed biological perspective to soil management in order to address longerterm sustainability challenges for crop production. A biological perspective is also critical to expanding soil assessment and management to address concerns over biodiversity, water quality, climate, recreation, and human and planetary health beyond humans.

The historical uneasiness with which scientists have embraced the concept of soil health is due to the challenges of defining soil health in a way that allows for a universal quantitative assessment that encompasses all of its ecosystem services, including human health. Reasons for this challenge include soil heterogeneity, the sitespecific nature of soil management and the varying ecosystem services that have sometimes conflicting or competing needs. Nevertheless, there has been widespread interest amongst researchers, policymakers and stakeholders in the use of the soilhealth concept.

In this perspective, we describe the relationship between soilhealth management and sustainable plant production, water quality, human health and climatechange mitigation is described. Biological, chemical and physical indicators and their integration into a comprehensive approach to soil health are outlined, and we argue for a greater inclusion of biological indicators in soilhealth assessments. Finally, recent technology developments that should be leveraged in measuring and monitoring soil health, and future directions for soil health research and management is discussed.

The increasing broad public interest in the soil-health concept is largely grounded in historical development. Even though the term 'soil health' has been more regularly used in the scientific and popular literature only since the early 2000s, the analogy of the soil ecosystem to an organism reaches far into the past. Soil is frequently part of creation myths and humans have always had deep spiritual connections with soil, as shown in songs, fine and performing arts.

Since the 1700s, scientists have introduced the notion of biological processes in the formation of soil and that soil ecosystems are endangered as much as any other ecosystem provided a foundation for soil health. The 1979 Gaia concept popularized the view of nature as a planetary-scale self-regulation system, explicitly including soil-ecosystem concepts and going beyond soil services solely for humans. Appreciation of soil biological processes has been largely enabled by significant advances in analytical capabilities since the 1980s, including global mapping of soil Biodiversity during the 2010s. the formulation of the united Nations' sustainable Development Goals in 2015 provided a need to align soil functions with sustainability, which makes soil health a suitable platform. The soil-health concept emerged from soil quality in the 1990s and was initially met with considerable criticism. More recently, policymakers have embraced the concept, exemplified by India distributing soil health cards to 100 million farmers and major companies starting programs on soil health to manage their supply chains more sustainably including carbon sequestration in soils as a main approach in the United Nations Framework Convention on Climate Change (UNFCCC) process to withdraw atmospheric carbon dioxide enhanced the political urgency to implement suitable soil-health practices on a global scale. The rapid adoption of the soil-health concept after 2010 could partly be rooted in its flexibility and, thereby, the ability of different stakeholders to use it in their own way.

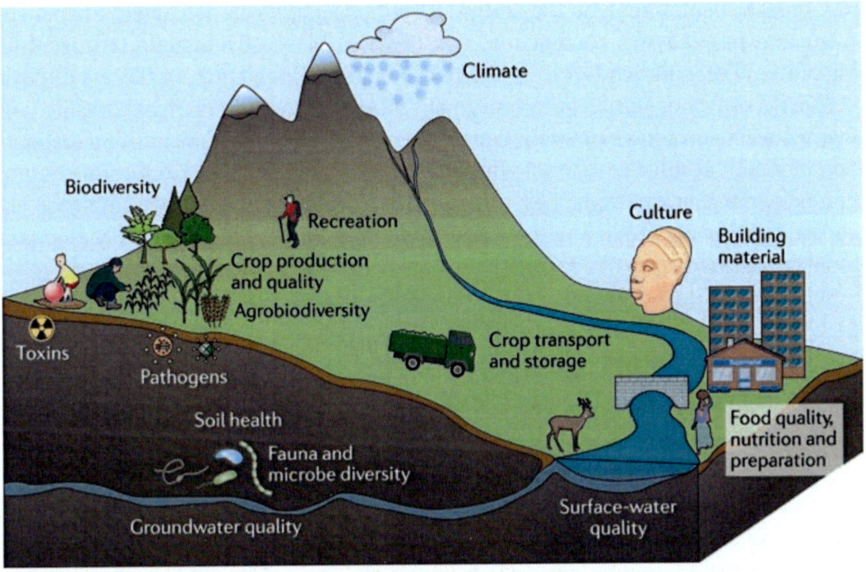

Fig. 6.2 Soil-health and global-ecosystem services. Soil health affects human and planetary health through crop production, quality, storage and transportation; food quality and taste; soil contamination; or through climate change, recreation and culture. Immediacy of soil-health effects on plants and soil biota facilitates assessment of causality (for example, soil nutrient availability affects crop production). Cascading effects (such as soil nutrient availability affecting human health indirectly through crop quality and food storage) require causalities to be demonstrated, for which, in some cases, the science still needs to be established, dimensional

Soil-Health and Soil-Ecosystem Services Soils provide multiple ecosystem services (Fig. 6.2) and, as such, soilhealth management in support of sustainability must consider three points: that enhancing many soilecosystem services requires multifunctional management; that managing soil to improve one service can have positive (synergistic) or negative effects (tradeoffs) on another service; and that soil-health management should sustain soil services over the long term. Here, we briefly highlight four main soilecosystem services—sustainable plant production, water quality control, human health advancement and climate change mitigation—that are considered during soil health management.

Sustainable Plant Production Plant production, the main goal of intensive agriculture, is an important focus of soilhealth management, as it affects water use and quality, human health, animal health, climate and biodiversity (Fig. 6.2). A foundation of soil health, though, is the recognition that managing nutrient availability alone, such as through the use of agrochemicals (mainly fertilizers), is not sufficient for optimizing plant growth. Furthermore, there is increased recognition that some management practices used in intensive agriculture to increase total plant production are detrimental to soil health. For example, rooting depth—critical in plant production depends, to a large extent, on soil structure, which is determined, in part, by organic matter content and soil preparation. Tillage can negatively impact soil structure through soil compaction, and the use solely of inorganic fertilizers (as opposed to organicrich fertilizers such as compost and manure, or the use of cover crops) is often not sufficient to restore or retain adequate levels of soil organic matter. Focusing on soil health will, therefore, expand soil management from a reliance on inorganic fertilizers to employing organic amendments and crop residue return, reducing mechanical impact by tillage, increasing plant diversity in both time and space or reducing erosion with contour ploughing (ploughing along elevation contours) or grass strips.

In addition to managing physicochemical soil properties for plant production, soil health considers the interactions between plants and soil microbial communities around roots, which can promote or reduce plant growth. Promoting a soil microbiome for high plant production requires management of microbial abundance and activity, community composition and specific functions. For example, organic amendments (such as compost) can foster increased resilience to plant pathogens through promotion of beneficial microorganisms. In many cases, higher organic matter content through higher amendments or reduced tillage increases biodiversity, which is expected to improve crop resilience. However, there are exceptions to these trends—reducing tillage, for example, can reduce crop yields in some instances, with followon reductions of soil organic carbon.

Water Quality Soil can be a source and/or sink of pollutants as rainwater and snowmelt move through it (Fig. 6.2). These pollutants include herbicides, pesticides, heavy metals, antibiotics, hormones, microplastics, pathogens, polycyclic aromatic hydrocarbons, perfluoroalkyl and polyfluoroalkyl substances. Moreover, nutrient pollution from agricultural fertilizer use is a global problem, leading to eutrophication and/or anoxia of waterways, promoting harmful algal blooms and negatively impacting drinking water quality. Thus, there is a tradeoff between soil management to support crop growth and water quality, which requires careful consideration and multiple management strategies.

Managing soil health to promote good water quality includes retaining pollutants and others in the soil, buffering against them and biotically transforming them. Increasing soil organic matter will retain heavy metals and organic toxins, some of which show nearly irreversible adsorption to organic matter. Using buffer zones, such as vegetative filter strips near agricultural areas or constructed wetlands, can slow the migration of nitrate, phosphate or pesticide contamination to water. Soil biota can transform organic pollutants, such as the common hydrocarbon toluene, into harmless compounds. Therefore, both organic matter content and microbial activity, key properties of soil health, improve the quality of the water that is draining soil.

Soil health of urban soils has not yet received sufficient recognition but can contain an even wider range of contaminants than agricultural soils, and many urban soils have also been modified to an extent that water can drain either very quickly or not at all. Soil health management in urban soils must, therefore, balance eliminating surface runoff against retaining water and pollutants by reduced drainage. A combination of managing physical retention with biological transformation of pollutants through high soil biodiversity is the goal of bioretention and constructed soils to provide clean drinking water.

Human Health Human health depends, to a great extent, on soil health, including and going beyond the obvious connection between soil and human health through crop production (Fig. 6.2). Similarly important is the type of crop and its nutritional content; soils with greater micronutrient availability are related to lower malnutrition and higher soil organic matter improves the nutritional value of crops. In addition to these relatively wellknown properties, the nutritional value of crops can also depend on robust soil biodiversity, which can enhance micronutrient bioavailability to crops and suppress soilborne plant disease, as well as affecting taste, food storage and food preparation.

Soils can also negatively impact human health. For example, soil pollutants can contaminate produce through direct contact or dust, suspension or rain splash. Some compounds, such as arsenic and most inorganic pollutants, can also be taken up through the root system and accumulate in grain or fruit. In addition to abiotic contaminants, soils can contain pathogenic fungi that produce mycotoxins, contaminating plant products and causing acute and chronic diseases in animals and humans. Furthermore, soils are also the source of parasitic worms (helminthiasis) that can live for years in the human gastrointestinal tract, cause malnutrition and result in stunted development.

Although soil hosts pathogens, it has also historically been the source of organisms that produce antibiotics used in the medical industry, such as streptomycin. Most of the soil microbiome remains to be identified, and important discoveries for human medical applications could still be made. Quantifying and managing soil biodiversity, part of the goals of soilhealth management, is needed to arrest extinction of microbial species and preserve opportunities for future bioprospecting.

Climate Change Intelligent soil management can mitigate or exacerbate climate change and its effects on other soilecosystem services, such as water quality or plant production (Nair 2019a, b).

For example, climatechangemitigation strategies, such as sequestering carbon in soil as organic matter, can benefit agriculture by improving crop productivity and resilience to drought and flooding. Furthermore, increased soil organic matter can be achieved by increasing the use of organic fertilizers or soil amendments, as well as by reducing tillage, to increase aggregation and control microbial mineralization to carbon dioxide (Table 6.1), which can also promote plant growth. However, there are tradeoffs between managing soil which are commonly used to increase crop production, which can lead to increased emissions of nitrous oxide, which is a powerful greenhouse gas. It is also very important to note that unbridled use of urea fertilizer, as was the case with green revolution in India, especially in Punjab state, can lead to enormous emission of nitrous oxide (N2O), which will enhance global warming, and, it has been noted that as much as 35% of global warming is contributed by N2O emission (Nair 2019b). These examples highlight the difficulty in balancing the various uses of soils and why it is important to provide context and goals for soilhealth management.

6.4.1 *Quantifying Soil Health*

Quantification is important in managing soilhealth and soilecosystem services, and the multifunctionality (Fig. 6.2) and diversity of soil requires multiple indicators to be quantified and integrated into an index.

Broadly, soilhealth indicators can be classified as physical, chemical or biological, although these categories are not always clearly delineated, as many properties are a reflection of multiple processes. For example, soil aggregation is the result of chemical parameters (such as organic matter content), mineral type and/or biological activities. Similarly, plantavailable phosphate falls under chemical indicators, but is largely a result of biological processes of microbial mineralization and plant uptake. The present classification (chemical, physical, biological) is, therefore, in many respects, less a reflection of causality (for example, as plant availability of phosphate is also a result of a biological process) than the object of enquiry (for instance, phosphate is a chemical indicator) that can be readily analysed.

To be used as a soilhealth indicator, a parameter should satisfy several criteria, which include being: relevant to soil health, its ecosystem functions and services (Table 6.1; Fig. 6.3); sensitive, by changing detectably and quickly without being reflective of merely shortterm oscillations; practical, by being conducted cheaply and with a short turnaround time; and informative for management (Fig. 6.4). Approximately half of the indicators currently used in more than 20% of 65 soilhealth analysis schemes (comprising a mixture of those declaring themselves to be soilquality or soilhealth schemes) satisfy all four criteria (Fig. 6.4), but some important indicators do not. Total organic carbon, for example, satisfies three criteria, but typically does not change very quickly (is not sensitive), requiring additional indicators, such as organic carbon fractions, that are more sensitive. Other indicators, such as soil texture or depth, do not readily change, cannot be easily managed (in other words, are not 'informative', Fig. 6.4) even though they are highly relevant for soil health and, in many schemes, still require timeintensive analyses or infield measurements.

However, these unmanageable indicators provide context for soil health and can be understood as mapping a soil's potential or capability, without which the manageable attributes cannot be understood. Importantly, and problematically, none of the listed biological indicators is currently effective in allowing cheap, reliable and quick information to be obtained.

Soilhealth assessments for plant production often include total organic carbon, plantavailable nutrients, pH, health for climate change versus for food production. For instance, the use of nitrogen fertilizers, cationexchange capacity, electrical conductivity, penetration resistance, nitrogen mineralization and microbial biomass

Table 6.1 Soil indicators, inclusion in ecosystem-service assessments, indicator type and assessment methods

Indicator	Inclusion[a]	Ecosystem service				Type of indicator	Methods to assess[b]
		Plant production	Water quality	Human health	Climate control		
Nitrogen–/sulfur–/phosphorus-mineralizing enzyme activity	<20%	+	+	–	+	B	Colorimetry, extraction; lab-on-a-chip; electrochemistry
Nitrogen mineralization	>20%	+	+	–	+	B	Incubation; extractions; lab-on-a-chip; electrochemistry
Microbial biomass	>20%	+	+	–	+	B	Incubation; extractions; lab-on-a-chip; electrochemistry
Pathogens	Proposed	+	+	+	–	B	Extractions; optical analyses; lab-on-a-chip; colour reactions; DNA probes; electrochemistry
Biodiversity	Proposed	+	+	+	+	B	Extractions; bioassays; metagenomics; high-throughput sequencing; phospholipid fatty acid; lab-on-a-chip
Microbial activity	>20%	+	+	+	+	B	Incubation; lab-on-a-chip; electrochemistry; biosensors
Parasites	Proposed	–	–	+	–	B	Extractions; bioassays; metagenomics; high-throughput sequencing; screening for pathogenicity genes; lab-on-a-chip; electrochemistry; ultrasound
Fauna	Proposed	+	+	+	+	B	Extractions; bioassays; metagenomics; high-throughput sequencing; lab-on-a-chip; electrochemistry; sound
Earthworms	<20%	+	–	+	–	B	Extractions; lab-on-a-chip; sound
GHG emissions	Proposed	–	–	–	+	B	In-field and laboratory GHG sensors; robots; lab-on-a-chip; biosensors
Organic toxins	Proposed	+	+	+	–	C	Extractions; passive samplers; lab-on-a-chip; electrochemistry
Organic chemical fractions	<20%	+	+	–	+	C	Near–/mid-infrared spectroscopy; density and size fractionation; oxidation
Organic nitrogen fractions	<20%	+	+	–	+	C	Protein assay; near–/mid-infrared spectroscopy; density and size fractionation
Organic carbon	>20%	+	+	+	+	C	Near–/mid-infrared spectroscopy; combustion; ultrasound
Bioavailable nutrients	>20%	+	+	+	+	C	Near–/mid-infrared spectroscopy; extractions; passive samplers; colorimetry; electrochemistry

(continued)

Table 6.1 (continued)

Indicator	Inclusion[a]	Ecosystem service				Type of indicator	Methods to assess[b]
		Plant production	Water quality	Human health	Climate control		
pH	>20%	+	+	+	+	C	Near–/mid-infrared spectroscopy; extractions; passive samplers; colorimetry; electrochemistry
Cation-exchange capacity	>20%	+	+	–	–	C	Near–/mid-infrared spectroscopy; extractions; passive samplers; colorimetry; electrochemistry
Electrical conductivity	>20%	+	+	+	–	C	Near–/mid-infrared spectroscopy; extractions; passive samplers; colorimetry; electrochemistry
Compound diversity	Proposed	–	+	–	+	C	Spectroscopy
Mobile nutrients	>20%	–	+	–	+	C	Near–/mid-infrared spectroscopy; extractions; passive samplers; colorimetry; electrochemistry
Heavy-metal toxins	>20%	+	+	+	–	C	Near–/mid-infrared spectroscopy; extractions; passive samplers; bioassays; lab-on-a-chip; biosensors; electrochemistry
Pore-size diversity	Proposed	–	+	–	+	P	Near–/mid-infrared spectroscopy; ultrasound
Aggregation	<20%	+	+	–	+	P	Sieving; near–/mid-infrared spectroscopy; ultrasound; visible imaging; infiltrometry
Water storage	<20%	+	+	+	+	P	Near–/mid-infrared spectroscopy; pressure plate
Penetration resistance	>20%	+	+	–	+	P	Penetrometry; mid-infrared spectroscopy
Infiltration	<20%	+	+	+	+	P	Near–/mid-infrared spectroscopy; ultrasound; visible imaging; infiltrometry

Indicators included in more than 20% of soil-health assessments are labelled as '>20%'. Those included in at least one but fewer than 20% of assessment methods are labelled as '<20%'. Those that are typically not included, but recommended to be included, are labelled as 'proposed'. Those indicators less directly relevant for a certain ecosystem service are marked as '–', while those that are more relevant are marked with '+'. *B* biological, *C* chemical, *GHG* greenhouse gas, *P* physical

[a]Proportion from reference.[6]

[b]Broad categories are given; for some, detailed methods have been proposed (see reference), while others are suggestions for future exploration

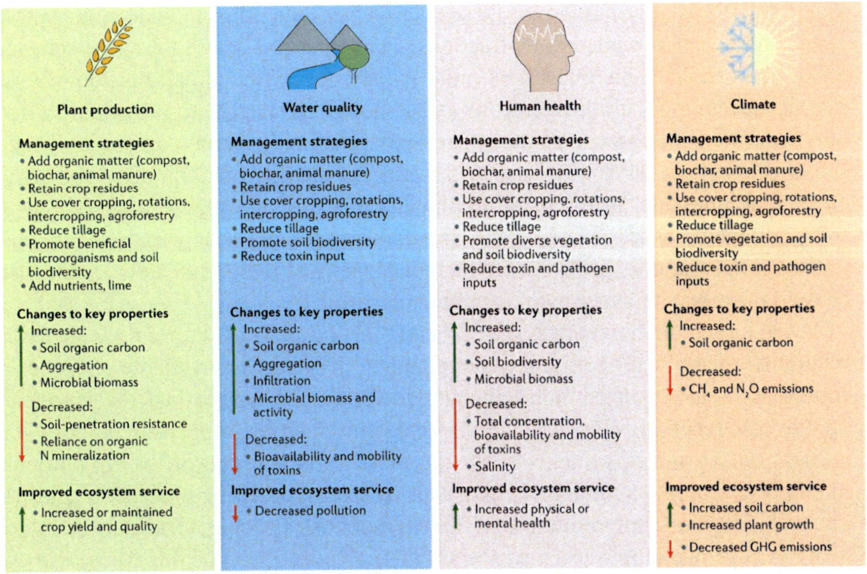

Fig. 6.3 Soil-ecosystem-services management. Four important roles of soil (plant production, water quality, human health and climate mitigation) are listed at the top of the figure. Various management strategies, and their impacts on key soil properties and ecosystem services, are listed underneath. *GHG* greenhouse gas, *N* nitrogen

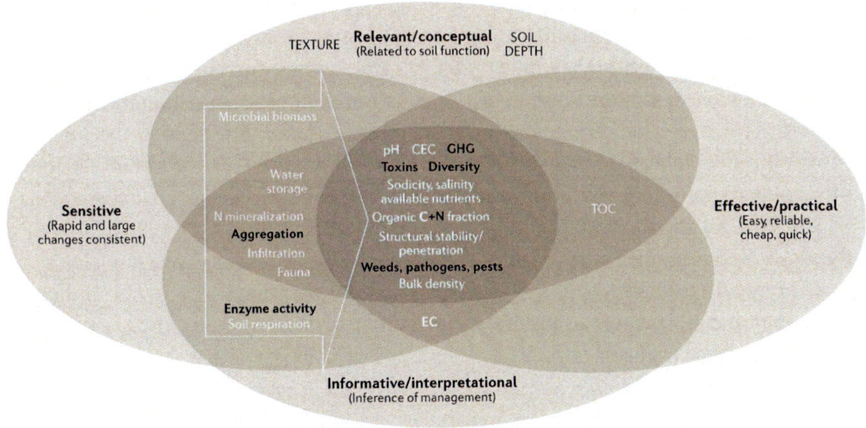

Fig. 6.4 Soil-health indicators and relevance to assessments. Soil-health indicators ideally are informative, sensitive, effective and relevant. Some do not fulfil all criteria but are still relevant (such as texture or soil depth, which do not change readily and are not managed and, therefore, also called capability indicators). Bold text denotes indicators that expand the utility of soil-health quantification beyond crop production towards sustainability and planetary health; the white arrow outline encompasses indicators that should be further developed to be effective and practical. Note, diversity includes biota in soil, diversity of soil types in landscape, molecular/structural diversity in soil organic matter and plants growing in soil, some of which might not be readily quantified through analytical or modelling approaches. *C* carbon, *CEC* cation-exchange capacity, *EC* electrical conductivity, *GHG* greenhouse gases, *N* nitrogen, *TOC* total organic carbon

(Table 6.1). A smaller number of these tests (fewer than 20%) include aggregation, water storage and organiccarbon fractions. Managing soil health for climatechange mitigation should include testing similar parameters, with a small portion of tests already examining soil nitrogen forms that should be adapted to provide information about potential greenhouse gas emissions, including nitrous oxide. Soilhealth assessments relevant for water quality should include microbial biomass and activity, mobile nutrients, heavy metal toxins and total organic carbon already part of many soilhealth testing schemes, yet, should also encompass aggregation and infiltration, which are only occasionally included. Many of these indicators should also be used in soilhealth assessment for human health.

In total, more than twothirds of soil health test frameworks currently include the traditional quantification of soil organic matter, pH and plantavailable phosphorus and potassium, and more than half include water storage and bulk density. Nair (2019a) has recently been arguing and advocating the precise measurement of soil nutrient buffer power, principally for the most important and diffusible nutrients like Phosphorus, Potassium and Zinc, and, its role in precise soil tests for nutrient bioavailability in crop production in three continents, Europe, Africa and Asia, for many field grown crops, as a very important indicator of soil health. These aspects are separately discussed in a book titled Intelligent Soil Management For Sustainable Agriculture The Nutrient Buffer Power Concept (Nair 2019a). A third of tests also recommend measurements of soil respiration, microbial biomass or nitrogen mineralization to characterize biological properties, as well as structural stability, Chemical indicators make up at least 40% of the indicators in 90% of the soilhealth-assessment schemes (Fig. 6.5), underscoring the continued importance of chemical properties in soilhealth quantification and the longstanding emphasis on plant production. Indeed, the most advanced analytical schemes currently, such as the Soil Management Assessment Framework, focus on indicators for sustainable crop production, However, the European Commission recently recommended the inclusion of soil biodiversity as one of six indicators of soil health. It is important to note in this context that "The Nutrient Buffer Power Concept" has been accepted as a benchmark for the ambitious multi-million US dollar funded project (USAID), of the Soils Department, Kansas State University, for soil testing in West Africa.

Biological indicators typically still constitute fewer than 20% of the indicators (Figs. 6.4 and 6.5), even when the total number of indicators used by a particular scheme increases. Furthermore, the development of soilhealthassessment schemes over the past decade has not yet led to inclusion of a greater proportion of biological indicators, despite their declared importance for soilhealth management (Fig. 6.5). One reason for the low representation of biological indices is, we posit, the lack of mechanistic understanding of how soil biota relate to soil functions (meeting the 'relevant' criteria, Fig. 6.4), how that understanding relates to management decisions ('informative') and the inability to easily quantify biological indicators ('effective'). This lack of understanding is even the case for soilecosystem services that would benefit from biological indicators, such water quality or biodiversity. In a Swiss grassland soil, for example, a decrease in soil biodiversity (microbes and fauna) was associated with lower plant diversity, a threefold higher phosphorus

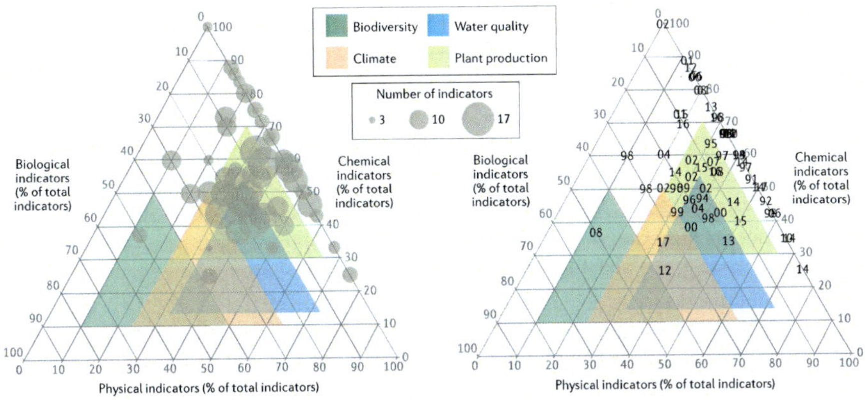

Fig. 6.5 Biological, chemical and physical indicators included in soil-health-assessment schemes. The left panel shows the number of indicators and proportion of each type (biological, chemical or physical). Each circle represents one assessment scheme, and the size of the circle represents the number of indicators in the scheme. The right panel shows the year of each soil-health-assessment scheme from the left panel. Only the last two digits of the year are shown (values in the 80s and 90s are from the 1980s and 1990s, and values from 00 to 20 represent years 2000 onwards). Currently proposed soil-health indices utilize mostly chemical and physical indicators. The proportion of biological indicators is typically lower than either chemical or physical indicators, which did not change over time as the methods were published, likely reflecting the historic focus of soil-health indices on crop growth. The number of indicators in the proposed schemes does not relate to the proportion of biological indicators. A comprehensive soil-health index might consider a balanced set of indicators that represents at least 20% biological, physical and chemical measurements. However, indices designed to quantify different services could require a different set of indicators: a soil-health index for plant production might require more chemical indicators (inside the yellow triangle), for water quality more physical (blue triangle), for biodiversity more biological (green triangle) and for climate more physical and biological indicators (orange triangle)

leaching and sixfold higher gaseous losses of nitrous oxide. Advancing both the information about causality between biological indicators and soil health, and those assessment tools that satisfy all four criteria, is, therefore, critically needed and is the next frontier in soilhealth research.

6.4.2 A New Generation of Indicators

Each soilhealth goal requires a different set of parameters be monitored, compared with reference states when appropriate and managed. For indicators included in more than 20% of alreadyproposed methods, we recommend that these be the minimum set of indicators for that management goal (Table 6.1). Furthermore, we suggest that additional measurements, especially biological assessments, be added when assessing soil for each of the management goals. For example, we suggest that aggregation, infiltration, earthworm abundance and organic C as crop production, and N fractions should be more widely adopted in soilhealth testing (Table 6.1), and Nmineralizing enzyme activity be added for soilhealth assessments for plant production. We further propose that several new indicators that are mainly geared towards nonagricultural soil services, such as human health and water quality, need to become part of routine soilhealth testing. These indicators include pathogens, parasites, biodiversity, bioavailable and mobile toxins (such as dioxin, polycyclic aromatic hydrocarbons and microplastics), and compound and poresize diversity.

Importantly, development of soilhealth indicators related to the climatechange functions of soils, such as greenhouse gas emissions and carbon sequestration, has largely been ignored (Nair 2019b). This neglect is largely due to greenhouse gas emissions depending on fluctuating conditions (such as moisture and temperature), so the magnitude of greenhouse gas fluxes for a given field or region cannot be assessed by onetime soil measurements. However, soil carbon fractions of both unprotected and mineralprotected organic matter already allow assessment of soil organic matter vulnerability with respect to soil carbon sequestration, and are indispensable indicators for soil's climatechange function. Such fractions capture changes in soil organic matter properties very sensitively, yet are less variable than and N fractions should be more widely adopted in soilhealth testing (Table 6.1), and Nmineralizing enzyme activity be added for soilhealth assessments for plant production. We further propose that several new indicators that are mainly geared towards nonagricultural soil services, such as human health and water quality, need to become part of routine soilhealth testing. These indicators include pathogens, parasites, biodiversity, bioavailable and mobile toxins (such as dioxin, polycyclic aromatic hydrocarbons and microplastics), and compound and poresize diversity.

Importantly, development of soilhealth indicators related to the climatechange functions of soils, such as greenhouse gas emissions and carbon sequestration, has largely been ignored. This neglect is largely due to greenhouse gas emissions depending on fluctuating conditions (such as moisture and temperature), so the magnitude of greenhouse gas fluxes for a given field or region cannot be assessed by onetime soil measurements. However, soil carbon fractions of both unprotected and mineralprotected organic matter already allow assessment of soil organic matter vulnerability with respect to soil carbon sequestration, and are indispensable indicators for soil's climatechange function. Such fractions capture changes in soil organic matter properties very sensitively, yet are less variable than mineralization or microbial biomass assays, allow unambiguous interpretation and can be quantified using rapid infrared technology (Table 6.1). Infield methods for measuring greenhouse gas emissions will need to provide integrated information about the highly temporally dynamic processes, requiring a new generation of sensors based on autonomous gas and solute detection powered by bioreactors and a range of energyharvesting technologies in wireless networks.

Diversity indicators, whether organismal (biological), molecular (chemical) or structural (physical), are not adequately included in or integrated into analytical frameworks of soil health. Biological diversity in particular has been recognized as important for soil and human health, yet, appropriate soilhealth indicators and practical quantification methods for soil biota diversity are lacking. Similarly, molecular and soil structural diversity are not yet explored but are important for soil organic carbon persistence and sequestration Nextgeneration sensor technology for plant and climate functions could provide the muchneeded platform to monitor changes in soil health over time. Recent global mapping of biodiversity and similar efforts will potentially provide context and reference sites for biodiversity calibration. Rapid screening techniques using nearinfrared and midinfrared, beyondinfrared energies, sound, lab onachip technology—technologies generally underdeveloped for soil —should be adapted to make existing soilhealth analyses cheaper and faster.

Further promising tools or techniques for observing biological properties, including electrochemistry and biosensors, are promising avenues that speak to the rapid emergence of new approaches. Similarly, passive samplers can and should be used to quantify the small proportion of organic toxins that are harmful to organisms, rather than assessments relying on total content that are not sufficiently sensitive to changes in management nor reflect the ecologically relevant fraction. Altogether, such technologies could expand the suite of assessed biological properties to include soil organic matter vulnerability and microbial or faunal community or functional gene information.

Advances in soilhealth monitoring over the coming decade should also include development of remotesensing techniques. Remote sensing should not only include spatial information of soil properties, such as seen with successes measuring soil moisture using microwave, but also assess soilmanagement practices that can be related to soil functions via mathematical modelling, as is already in development for soil organic carbon monitoring. Such rapid and largescale soilhealth screening through remote sensing should be complemented by exploring the use of guided smallscale robotics to assess soil hotspots and sensitive flowpaths (such as soil cracks and earthworm holes) that are typically undetected through remote or bulk assessments. Nextgeneration electronics should be applied to enable cheap and distributed sensor deployment, fast data transmission, storage and handling, and need to make use of the rapid development in the computing and smartgrid sector to develop internetofthings sensor networks for soilhealth monitoring. Rapid screening and *in situ* and remote monitoring technologies discussed here would substantially advance our ability to measure and manage soil health, ultimately improving soilecosystem services.

6.4.3 Soil-Health Indices

As there is a multitude of soilhealth indicators, an appropriate desire exists among scientists and stakeholders to integrate them into one single test score or 'soilhealth index' (note the difference between 'indicator' and 'index'). However, relatively few indices exist; in the 2020 database compiled on soil health, Soil Health DB, which assessed over 500 studies on soil health and quality, only five studies included a single soilhealth index. We discuss some of the challenges in creating integrated indices and needs that must be overcome when developing and using them.

Challenges Creating a soilhealth index is difficult, as indicated by the relatively low number of published indices, because it requires quantitative transformation and weighting of multiple indicators, including categorical properties, in order to integrate them into a final single score. Indicator values are necessarily transformed using non-linear relationships, because a higher value does not always indicate better soil health. A ‘high’ organic carbon value might indeed indicate a desirable property for many soil functions, but pH should be within an intermediate range, and the force needed to penetrate the soil should be relatively low. In the Comprehensive Assessment of Soil Health, for example, these three categories are described as ‘More is better’, ‘Optimum curve’ and ‘Less is better. In most existing frameworks, the conversion of measured values to scores is based on the distribution of the actual measurements within a reference dataset. To determine the final soilhealth score, often, all indicators are treated as equally important. For instance, the Comprehensive Assessment of Soil Health assigns values between 0 and 100 (where 100 is the highest) to each indicator based on a comparison with reference values of all available data in the region.

Although these indices can provide useful information on large scales, regional comparisons are not appropriate in situations with bias resulting from inherent differences between soil types and require careful calibration to regional conditions and needs. In temperate arable soils in England and Wales, for example, an organic carbon value of 1.5% is considered a lower limit for soils with 40% clay, but would be considered high in soils that have less than 10% clay. Therefore, identifying soil organic carbon as high or low in this region depends on clay content, and soils should be compared with references with a similar clay content.

Changes of soil health over time can generate more robust comparisons, which relates to the definition of soil health as a ‘continued capacity’. For instance, the formation or maintenance of aggregates over time can indicate better soil health, as particles are bound into aggregates mainly by microbial products from organic amendments. However, aggregates can form even without organic matter, and the formation of aggregates differs between soils—within weeks and without organic amendments, aggregates formed in a kaolinitic Oxisol from Brazil, whereas no aggregates formed in an illitic Mollisol from the USA. Considering inherent differences between soils is particularly important when using biological indices. In one example, bacterial diversity was as much affected by soil type, soil texture and pH as by whether soils were located under forests or grasslands across a north–south gradient in Germany. At the same sites, changes in bacterial diversity as a result of fertilization, mowing and grazing in grasslands or of various silvicultural management in forests were only discernible within a given site.

Despite these caveats, appropriately comparing changes in soilhealth indicators and indices over time or with a suitable reference dataset can be used to assess whether, for example, a reduction in tillage or addition of compost improves aggregation and total soilhealth scores. Indeed, it is standard practice to identify whether a soil has high or low amounts of extractable nutrients or to convert nutrient indicators into amounts of fertilizer to apply to a certain crop while recognizing differences in texture and mineral types, which even utilizes information from fertilizer responses for a specific soil. These aspects are elaborately discussed in the book “Intelligent Soil Management For Sustainable Agriculture The Nutrient Buffer Power Concept”, Springer (Nair 2019a).

Needs Development of a soilhealth index that includes all soil functions (Figs. 6.1 and 6.2) requires engagement of a broader set of stakeholders than an index focused on crop production. A comprehensive soilhealth framework will need to include and allow weighting tradeoffs to lead to optimum overall function, as it must balance the sometimes competing functions of soil, for example, the need to minimize water pollution by fertilizers versus the need to optimize nutrient availability for crop growth. Such tradeoffs also mean that the effects of noncropecosystem services such as water quality have to be valued against cropgrowth effects on human health, which has rarely been done in a quantitative way, even in comprehensive ecosystemservices assessments. Soil effects on human health need to be assessed as they affect humans both through production of nutritious food and through clean water, with unclear quantitative criteria on whether water is more important than food or *vice versa*.

Holistic soilhealth indices should, therefore, include multi-criteria decision analysis to quantify and prioritize sustainability outcomes of soilhealth management. Societal demands for different soil functions such as water quality and food production can vary by stakeholder and region. In an analysis of societal demands in Europe, water quality and food production was, on average, mentioned by the same groups, although densely populated countries such as the Netherlands and Belgium put more value on water quality and nutrient management than countries such as Romania or Finland. Soilhealth data should be presented using interactive data visualizations that reconfigure according to the desired focus. Such interactive tools will benefit researchers as well as stakeholders to prioritize soil functions and make decisions. Emerging dataanalysis tools such as machine learning, deep learning, artificial neural networks or game theory should be explored more fully in order to quantify the effect of soilhealth indicators as well as prioritize soil functions such as water quality or food production.

In parallel, new analytical and conceptual approaches need to be developed that capture systems characteristics of soil health, in order to operationalize both monitoring soil health itself and understanding soilhealth effects on soil functions. Avery interesting original approach, now globally known as **"The Nutrient Buffer Power Concept"** (Nair 1996, 2019a) assume considerable importance in this context. Precision and digital agriculture are expanding avenues to leverage for quantification of soil health with its multiple ecosystem functions and services. There must be greater engagement between soil science and engineering, whereby both instrument and computational technology is jointly developed with stakeholders. For example, soilengineering collaborations through colabs will need to advance scientific discovery of new detector technology as well as dataanalysis tools that can adapt complex data structures into simple apps for stakeholder use. Water science, medicine, psychology, philosophy and other fields need to engage for metrics and management to reflect the full range of soilhealth functions, including climate change, water quality, biodiversity and human health.

Fostering discussions at professional and trade meetings, as well as crosstraining of the next generation of scientists, will help to promote mutual understanding and joint problem solving.

6.4.4 Future Perspectives

The soilhealth concept fills an important stakeholder need in sustainable development by elevating the recognition of the role of soil in modern society and is developing into an attractive and actionable platform for farmers, land managers, municipalities and policymakers. The versatility of the concept allows many stakeholders to adopt soil health and to make it work for their context. By providing an illustrative link to broader sustainability goals that can motivate innovative soil management, soil health meets universal agreement in the eye of the public as a goal to work towards.

Scientists are converging on a definition of soil health and are developing or refining methods to quantify its various facets, albeit mainly with respect to its crop-productivity function and with inadequate consideration of biotic and abiotic diversity. Researchers should embrace soil health as an overarching principle to which to contribute knowledge, rather than as only a property to measure.

In this way, soil health could become better established as a scientific field to which many disciplines can contribute, for example, by listing their specific discipline's research also under the keyword 'soil health'. Making the soilhealth concept live up to its potential as a unifying concept that integrates soil functions requires engagement by all involved parties, and, particularly, a common understanding between stakeholders and scientists.

Because of soil's broad environmental and societal functions, soil health should be legally recognized as a common good. The development of soilhealth-quantification standards should be spearheaded by governmental or intergovernmental organizations such as the Global Soil Partnership. International standards have to be developed for suitable types of indicator, their methodological details and their integration into indices. Such a comprehensive soilhealth index should then be referenced by local, regional or national jurisdictions and organizations to guide decisions that impact soil and its functions to benefit sustainability goals.

Acknowledgments With love I dedicate this book to my wife, Pankajam, who sustains me in this very difficult journey that life is Addendum. "The Nutrient Buffer Power Concept" project was shortlisted for the very prestigious U.S. $1 million Rolex Awards for Enterprise 2012 for its originality from more than 3500 nominations worldwide and is the only project selected for this coveted distinction from the Asian continent.

References

Altieri MA (1983) Agroecology: the scientific basis of alternative agriculture. Division of Biological Control, University of California, Berkley

Altieri M (1995) Agroecology: the science of sustainable agriculture. Westview Press

Barres D, Bonny S, Le Pape Y, Remy J (1985) Une ethique de la pratique agricole. Agriculteurs biologiques du Nord-Drome, Paris. Grignon, Grenoble, France, INRA-ESR

Besson Y (2007) Histoire de lágriculture biologique: une introduction aux fondateurs, Sir Albert Howard, Rudolf Steiner, Hans Peter Rusch, Masanobu Fukouka. Universite de Technologie de Troyes, CREIDD. PhD

Buck D, Getz C, Guthman J (1997) From farm to table: the organic vegetable commodity chain of Northern California. Sociol Rural 37:3–20
Dalgaard T, Hutchings N, Porter J (2003) Agroecology, scaling and interdisciplinarity. Agric Ecosyst Environ 100:39–51
Francis C, Lieblein G, Gliessman S, Breland T, Creamer N, Harwood R, Salomonsson L, Helenius J, Rickerl D, Salvador R (2003) Agroecology: the ecology of food systems. J Sustain Agric 22:99–118
Gibbons M, Limoges C, Nowtny H, Schwartzman S, Scott P, Trow M (1994) The new production of knowledge: the dynamics of science and research in contemporary societies. Sage, London
Gliessman S (1990) Agroecology: researching the ecological basis for sustainable agriculture. Ecol Stud 78:3–10
Goodman D (2000) The changing bio-politics of the organic production, regulation, consumption. Agric Human Values 17:211–213
Guthman M (2000) An agro-ecological assessment of grower practices in California. Agric Human Values 17:257–276
Halberg N (2007) The impact of organic farming on food security in a regional and global perspective: challenges and prospects. Issues paper presented during the conference "Organic Agriculture and Food Security", 3–5 May, 2007, FAO, Rome, Italy
Hall A, Mogyorody V (2001) Organic farmers in Ontario: an examination of the conventionalization argument. Sociol Rural 41:389–422
Hill SB (1985) Redesigning the food system for sustainability. Alternatives 12:32–36
Hill SB, MacRae R (1995) Conceptual frameworks for the transition from conventional to sustainable agriculture. J Sustain Agric 7:81–87
Hodges R (1981) An agriculture for the future. In: Stonehouse B (ed) Biological husbandry: a scientific approach to organic farming. Butterworths, London, pp 1–14
Kaltoft P (1999) Values about nature in organic farming practice and knowledge. Sociol Rural 39:39–53
Langer V (2002) Changes in farm structure following conversion to organic farming in Denmark. Am J Altern Agric 72:75–81
Lehmann J, Bossio DA, Kogel-Knabner I, Billig MC (2019) The concept of soil health. On line Publication
Lockie S, Halpin D (2005) The conventionalization thesis reconsidered: structural and ideological transformation of Australian organic agriculture. Sociol Rural 45:284–307
MacMahon N (2005) Biodynamic farmers in Ireland, transforming society through purity, solitude and bearing witness? Sociol Rural 45:98–114
Michelsen J (2001) Recent development and political acceptance of organic farming in Europe. Sociol Rural 41:3–20
Nair KPP (1996) The buffering power of plant nutrients and effects on availability. Adv Agron 57:237–287
Nair KPP (2019a) Intelligent soil management for sustainable agriculture. The nutrient buffer power concept, vol 389. Springer
Nair KPP (2019b) Combating global warming. The role of crop wild relatives for food security, vol 120. Springer
Niggli U (1999) Holistic approaches in organic farming research and development. In: Zanoli R, Knell R (eds) Research methodologies in organic farming. Proceedings of the first SREN workshop on research methodologies in organic farming, FAO & FiBL, pp. 33–40
Padel S (2001) Conversion to organic farming: a typical example of a diffusion of an innovation? Sociol Rural 41
Piriou S (2002) L'institutionnalisation de l'agriculture biologique (1980–2000). Ecole Nationale Superieure Agronomique de Rennes, PhD thesis
Robin P, Aeschlimann J-P (2007) Elements de reflexion sur la formation agronomique et l'agriculture de demain. In: Robin P, Aeschlimann JP, Feller C (eds) Histoire et Agronomi entre ruptures et duree. IRD, Paris, pp 457–474

Salmona M (1994) Les paysans francais. Le travail, les metiers, la transmission des savoirs. Paris, L'Harmattan

Smith E, Marsden T (2004) Exploring the "limits to growth" in UK organics: beyond the statistical image. J Rural Stud 20:345–357

Tress B (2001) Converting to organic agriculture – Danish farmers'views and motivations. Geografisk Tidsskrift 101:131–143

Van den Werf H, Petit J (2002) Evaluation of the environmental impact of agriculture at the farm level: a comparison and analysis of 12 indicator-based methods. Agric Ecosyst Environ 93:131145

Verhoog H, Matze M, Lammerts E, Baars T (2003) The role of the concept of natural (naturalness) in organic farming. Agric Human Values 16:29–49

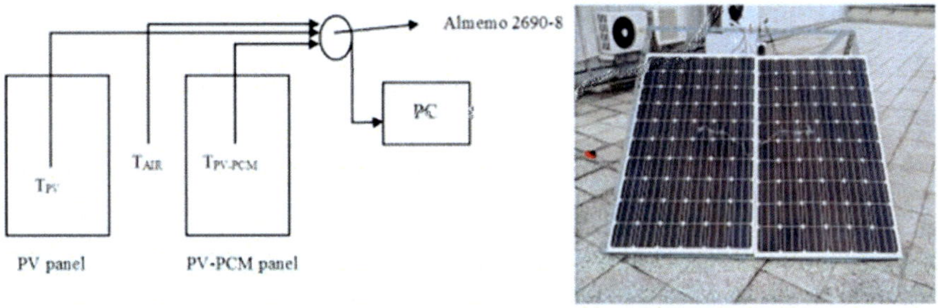

Fig. 5.18 Schematic diagram of experimental setup (left) and front side of PV and PV-PCM panel measurement (right) [174]

at back side of panel. The fins having 4-mm width were used for thermal analysis and system efficiency was studied along fins. The results indicated that by adding PCM, the temperature of panel was maintained to 40 °C for 80 min.

Haung et al. [177] investigated PCM-based PV/T system having metal fin configuration for cooling of PV system. The three PCMs, namely, RT27, RT35 and Waksol A were used by 85% to improve efficiency. It was observed that by using RT27 and fins, there was significant reduction in increment of temperature. Browne et al. [178] performed study on PCM-PV/T, PV/T-water and non-cooled PV panel. The stainless steel contained was used with tubing to pass water and PCM. The results concluded that PCM-PV system exhibited high heat storage compared to PV/T-water under same conditions. Browne et al. [179] also studied potential of PCM for efficient performance of PV panel, concentrated PV and building integrated PV panel. The authors work revealed that PCM-based system was a better technique for cooling system. Aktin et al. [180] studied four different techniques for thermal regulations of PV panels. The two monocrystalline silicon solar panels were under consideration. The four different configurations were investigated and the increment in overall efficiency of PV panel was 12.9% with last case having PCM and finned.

The efficiency of PV panels is still low by integrating with only PCM, which can be enhanced by various methods. One of the major techniques is to add NPs in PCM, which results in increment in thermal conductivity of PCM. The addition of high value of "k" of NPs enhances the performance capacity of base PCM. Al-Waeli et al. [181] investigated nano PCM and nanofluid-based photovoltaic thermal system to improve the overall efficiency of system. The SiC NPs were used in paraffin PCM and nanofluid (Water + SiC) at flow rate of 0.17 kg/s was used for improvement in value of "k" and to enhance thermal efficiency. The experiments were conducted under the outdoor conditions of Selangor, Malaysia. The results indicated that PV cell temperature was reduced by 30 °C, open circuit voltage enhanced from 11–13 to 20–21 V. The power was increased from 61.1 to 120.7 W, electrical efficiency improved from 7.1 to 13.7% and thermal efficiency reached up to 72%.

Al-Waeli et al. [182] also compared three different types of PV/T system with conventional PV system. The three various cooling techniques were proposed:

- Tank filled with water and water circulation in cooling pipes
- Tank filled with nano PCM (PCM + nano SiC)
- Tank filled with nanofluid (water + SiC) and circulation through cooling pipes.

The results indicated that nano PCM nanofluid-based PV/T system enhanced electric current from 3.69 to 4.04 A as well as electrical efficiency from 8.07 to 13.32% when compared with the conventional photovoltaic system. The three artificial neutral network methods were also implemented, and results agreed with experimental results. Al-Waeli et al. [183] carried out technical and economic study of PV/T system containing tank having nano PCM attached with PV panel, nanofluids inside the pipe, which passed through nano PCM tank at 0.175 kg/s as represented in Fig. 5.19. The results concluded that the efficiency of system was increased from 7.1 to 13.7% as compared to conventional system. The output water temperature was 39.52 °C and system was economically viable.

Dhaidan et al. [184] carried out tests experimentally as well as numerically to check the melting of nano PCM under constant heat flux. The value of "k" and melting rate was enhanced by adding NPs in base PCM. The rate of melting was higher at low value of NPs loading, which resulted in higher energy storage and saved cost in comparison with higher NPs loading. Groulx [185] investigated melting and freeing of nano PCM in a container having metallic fins in it. It was concluded that base pcm with metallic fins and NPs was performed much better in terms of thermal storage and HT rate.

Parameshwarah et al. [186] conducted the comparison of nano PCM with pure PCM in building applications for thermal storage. The nano PCM contained silver

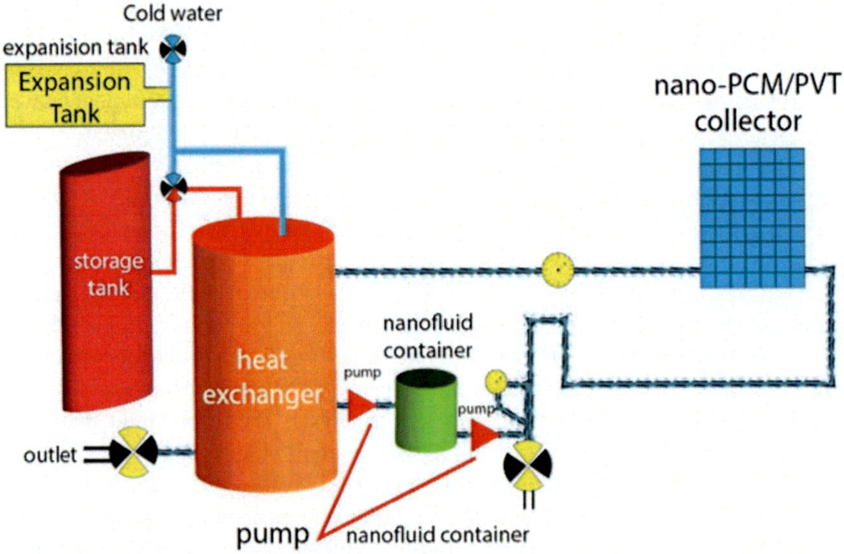

Fig. 5.19 Experimental setup proposed by Al-Waeli et al. [183]

and titanium NPs in base PCM and it was observed that freezing time of nano PCM was lower than pure PCM, which resulted in increment in value of "k" and thermal storage capacity. Amin et al. [187] enhanced the value of "k" of base PCM beeswax by adding graphene NPs, which was later used in building applications. It was observed that the value of "k" increased to 2.8 W/m.K and latent heat enhanced by 22.5% having 0.3 wt% of beeswax/graphene. Luo et al. [188] prepared form-stable composite PCM having mixture of capric acid, palmitic acid, steric acid and NPs of silica. It was observed that nano PCM having phase change temperature of 17.16–26 °C was suitable for maintaining indoor temperature of buildings. Sharma et al. [189] investigated the cooling effect of building integrated concentrated photovoltaic (BICPV) by using micro fins/PCM and nano PCM. The thermal efficiency was first with un-fins and with fins and then checked with PCM and nano PCM. The reduction in temperature in the center of system was 10.7 °C by using micro-fins with PCM and 12.5 °C by using micro-fins with nano PCM. The decrement in temperature without micro-fins of PCM and nano PCM was 9.6 and 11.2 °C respectively, compared to natural convection case. Ma et al. [190] presented nano PCM-based thermal storage in ceiling ventilation system integrated with PV/T system. The RT24 PCM was mixed with Cu NPs and conclusion was drawn. The results depicted that nano PCM has higher melting and solidification time was improved compared to pure PCM. The 8.3% of more heat was absorbed and 25.1% of more heat was released compared to pure PCM. Lin and Al-Kayiem [191] investigated the 20-nm five samples of copper nanoparticle dispersed into paraffin wax for solar energy thermal storage. The result demonstrated that for all dispersed weight of Cu nanoparticles, the thermal conductivity of Cu-PCM nanocomposite was increased (Table 5.4).

The chapter concludes the applications of advanced energy storage materials in a wide range of areas of thermal applications that include buildings, solar energy, waste heat recovery, seawater desalination, electronic cooling and photovoltaic thermal systems. These advanced materials have broadly used in these areas, which resulted in performance enhancement greatly. The researchers should continuously explore the properties of these materials to broaden their area of applications. There should be more areas of applications exploit in the field of thermal storage system for these materials for future research. More work is to be needed to explore the new types of energy storage materials for applications in various fields. The properties related to advanced storage materials should be considered further for utilization in more applications with much better execution.

Table 5.4 Summary of various authors on PV/T systems using energy storage materials

References	PCM	NPs	Nano concentration	Results
[192]	82 wt% co-conut oil and 18 wt% sunflower oil	Boehmite (AlO(OH))	0.009 w/w	Efficiency increased from 29 to 48% by nano-PCM at 690 W/m^2. Along 25 °C temperature was reduced when compared to without PCM
[193]	Paraffin (RT35HC)	Al_2O_3	0.11, 0.22, 0.33, 0.44, 0.55, 0.66, 0.77 (vol%)	The temperature was reduced from 1 to 4.5 °C from 0.11 to 0.77 vol% of NPs at 820 W/m^2
[194]	Coconut oil	CuO	0, 2, 4, 5 (vol%)	The melting temperature of nano PCM was improved
[195]	Paraffin	Cu	2 wt%	The 14.2% increment in thermal conductivity by adding NPs in base PCM
[196]	Paraffin	Alumina, Titania, silica	1, 2, 3 (wt%)	Titania showed more effective in terms of energy storage
[197]	Erythritol	Graphite particles	15 vol%	Thermal conductivity increased by 640% compared to pure PCM by adding graphite particles
[198]	Barium chloride	Titania particles	0.07, 0.13, 0.25, 0.50 (wt%)	The 12.76% enhancement in thermal conductivity at 0.50 wt% of Titania NPs was observed
[199]	Dedecanol	MWCNTs	1 and 2 (wt%)	The 4.6 and 11.0% enhancement in thermal conductivity by addition of 1 and 2 wt% of CNTs, respectively
[200]	Paraffin wax	CNTs	1, 2, 5, 10 (wt%)	The 5.9–24.4% increment in thermal conductivity by using CNTs in base PCM

(continued)

Table 5.4 (continued)

References	PCM	NPs	Nano concentration	Results
[201]	Paraffin wax	graphite	1, 4, 7, 10 (wt%)	The 4.3–167.5% enhancement in thermal conductivity of base PCM by adding graphite NPs
[202]	Paraffin wax	Al_2O_3, ZnO_2, SiC	0.1, 0.5, 1, 2, 3, 4 (wt%)	Thermal conductivity improved by 3.3, 1.8 and 4.2% with the addition of 1% nano-Al_2O_3, nano-ZnO_2 and nano-SiC, respectively
[203]	Paraffin wax	MWCNTs	0.2, 0.5% by mass	The maximum increment in value of "k" was observed 0.203 W/mK at 0.5% mass of MWCNTs

References

1. M. Sokolov, Y. Keizman, Performance indicators for solar pipes with phase change storage. Sol. Energy **47**, 339–346 (1991)
2. H.E.S. Fath, Technical assessment of solar thermal energy storage technologies. Renew. Energy **14**, 35–40 (1998)
3. E.-B.S. Mettawee, G.M.R. Assassa, Experimental study of a compact PCM solar collector. Energy **31**, 2958–2968 (2006)
4. A. Shukla, D. Buddhi, R.L. Sawhney, Solar water heaters with phase change material thermal energy storage medium: A review. Renew. Sustain. Energy Rev. **13**, 2119–2125 (2009)
5. M. Iten, S. Liu, A. Shukla, A review on the air-PCM-TES application for free cooling and heating in the buildings. Renew. Sustain. Energy Rev. **61**, 175–186 (2016)
6. E. Osterman, V. Butala, U. Stritih, PCM thermal storage system for 'free'heating and cooling of buildings. Energy Build. **106**, 125–133 (2015)
7. F. Guarino, A. Athienitis, M. Cellura, D. Bastien, PCM thermal storage design in buildings: experimental studies and applications to solaria in cold climates. Appl. Energy **185**, 95–106 (2017)
8. B. Nie, X. She, Z. Du, C. Xie, Y. Li, Z. He, Y. Ding, System performance and economic assessment of a thermal energy storage based air-conditioning unit for transport applications. Appl. Energy **251**, 113254 (2019)
9. S. Gharbi, S. Harmand, S. Ben Jabrallah, Parametric study on thermal performance of PCM heat sink used for electronic cooling, in *Exergy a Better Environment and Improved Sustainability*, vol. 1 (Springer, 2018), pp. 243–256
10. R. Kothari, P. Mahalkar, S.K. Sahu, S.I. Kundalwal, Experimental investigations on thermal performance of PCM based heat sink for passive cooling of electronic components, in *ASME 2018 16th International Conference on Nanochannels, Microchannels, Minichannels, American Society of Mechanical Engineers*, pp. V001T11A005–V001T11A005 (2018)
11. H. Usman, H.M. Ali, A. Arshad, M.J. Ashraf, S. Khushnood, M.M. Janjua, S.N. Kazi, An experimental study of PCM based finned and un-finned heat sinks for passive cooling of electronics. Heat Mass Transf. **54**, 3587–3598 (2018)

12. S.S. Chougule, V.V. Nirgude, S.P. Shewale, A.T. Pise, S.K. Sahu, H. Shah, Application of paraffin based nanocomposite in heat pipe module for electronic equipment cooling. Mater. Today Proc. **5**, 23333–23338 (2018)
13. A. Moldgy, R. Parameshwaran, Study on thermal energy storage properties of organic phase change material for waste heat recovery applications. Mater. Today Proc. **5**, 16840–16848 (2018)
14. K. Merlin, J. Soto, D. Delaunay, L. Traonvouez, Industrial waste heat recovery using an enhanced conductivity latent heat thermal energy storage. Appl. Energy **183**, 491–503 (2016)
15. A.I. Fernández, C. Barreneche, L. Miró, S. Brückner, L.F. Cabeza, Thermal energy storage (TES) systems using heat from waste, in *Advanced Thermal Energy Storage Technology* (Elsevier, 2015), pp. 479–492
16. M. Jaworski, R. Wnuk, M. Cieślak, B. Goetzendorf-Grabowska, Experimental investigation and mathematical modelling of thermal performance characteristics of textiles incorporating phase change materials (PCMs), in: *Environmental Engineering. Proceedings of the International Conference on Environmental Engineering* (ICEE, Vilnius Gediminas Technical University, Department of Construction Economics, 2017), pp. 1–9
17. X. Liu, Y. Lou, Preparation of microencapsulated phase change materials by the sol-gel process and its application on textiles. Fibres Text. East. Eur. (2015)
18. P. Zhang, J. Li, L. Lv, Y. Zhao, L. Qu, Vertically aligned graphene sheets membrane for highly efficient solar thermal generation of clean water. ACS Nano **11**, 5087–5093 (2017)
19. P. Zhang, Q. Liao, T. Zhang, H. Cheng, Y. Huang, C. Yang, C. Li, L. Jiang, L. Qu, High throughput of clean water excluding ions, organic media, and bacteria from defect-abundant graphene aerogel under sunlight. Nano Energy **46**, 415–422 (2018)
20. L. Zhou, Y. Tan, J. Wang, W. Xu, Y. Yuan, W. Cai, S. Zhu, J. Zhu, 3D self-assembly of aluminium nanoparticles for plasmon-enhanced solar desalination. Nat. Photonics **10**, 393 (2016)
21. X. Li, W. Xu, M. Tang, L. Zhou, B. Zhu, S. Zhu, J. Zhu, Graphene oxide-based efficient and scalable solar desalination under one sun with a confined 2D water path. Proc. Natl. Acad. Sci. **113**, 13953–13958 (2016)
22. K.-K. Liu, Q. Jiang, S. Tadepalli, R. Raliya, P. Biswas, R.R. Naik, S. Singamaneni, Wood–graphene oxide composite for highly efficient solar steam generation and desalination. ACS Appl. Mater. Interfaces **9**, 7675–7681 (2017)
23. H. Ren, M. Tang, B. Guan, K. Wang, J. Yang, F. Wang, M. Wang, J. Shan, Z. Chen, D. Wei, Hierarchical graphene foam for efficient omnidirectional solar–thermal energy conversion. Adv. Mater. **29**, 1702590 (2017)
24. J. Wang, Y. Li, L. Deng, N. Wei, Y. Weng, S. Dong, D. Qi, J. Qiu, X. Chen, T. Wu, High-performance photothermal conversion of narrow-bandgap Ti_2O_3 nanoparticles. Adv. Mater. **29**, 1603730 (2017)
25. Z. Yin, H. Wang, M. Jian, Y. Li, K. Xia, M. Zhang, C. Wang, Q. Wang, M. Ma, Q. Zheng, Extremely black vertically aligned carbon nanotube arrays for solar steam generation. ACS Appl. Mater. Interfaces **9**, 28596–28603 (2017)
26. T.R. Shah, H.M. Ali, Applications of hybrid nanofluids in solar energy, practical limitations and challenges: a critical review. Sol. Energy **183**, 173–203 (2019)
27. C. Chen, Y. Li, J. Song, Z. Yang, Y. Kuang, E. Hitz, C. Jia, A. Gong, F. Jiang, J.Y. Zhu, Highly flexible and efficient solar steam generation device. Adv. Mater. **29**, 1701756 (2017)
28. X. Hu, W. Xu, L. Zhou, Y. Tan, Y. Wang, S. Zhu, J. Zhu, Tailoring graphene oxide-based aerogels for efficient solar steam generation under one sun. Adv. Mater. **29**, 1604031 (2017)
29. L. Cui, P. Zhang, Y. Xiao, Y. Liang, H. Liang, Z. Cheng, L. Qu, High rate production of clean water based on the combined photo-electro-thermal effect of graphene architecture. Adv. Mater. **30**, 1706805 (2018)
30. C. Chen, L. Zhou, J. Yu, Y. Wang, S. Nie, S. Zhu, J. Zhu, Dual functional asymmetric plasmonic structures for solar water purification and pollution detection. Nano Energy **51**, 451–456 (2018)

31. V. Dao, H. Choi, Carbon-based sunlight absorbers in solar-driven steam generation devices. Glob. Chall. **2**, 1700094 (2018)
32. F. Zhao, X. Zhou, Y. Shi, X. Qian, M. Alexander, X. Zhao, S. Mendez, R. Yang, L. Qu, G. Yu, Highly efficient solar vapour generation via hierarchically nanostructured gels. Nat. Nanotechnol. **13**, 489 (2018)
33. X. Yang, Y. Yang, L. Fu, M. Zou, Z. Li, A. Cao, Q. Yuan, An ultrathin flexible 2D membrane based on single-walled nanotube–MoS2 hybrid film for high-performance solar steam generation. Adv. Funct. Mater. **28**, 1704505 (2018)
34. H.M. Ali, T.R. Shah, H. Babar, Z.A. Khan, Application of nanofluids for thermal management of photovoltaic modules: a review, in: *Microfluid. Nanofluidics, IntechOpen* (2018)
35. A. Abhat, Low temperature latent heat thermal energy storage: heat storage materials. Sol. Energy **30**, 313–332 (1983)
36. I. Dincer, S. Dost, X. Li, Performance analyses of sensible heat storage systems for thermal applications. Int. J. Energy Res. **21**, 1157–1171 (1997)
37. V.V. Tyagi, S.C. Kaushik, S.K. Tyagi, T. Akiyama, Development of phase change materials based microencapsulated technology for buildings: a review. Renew. Sustain. Energy Rev. **15**, 1373–1391 (2011)
38. M. Ravikumar, P.S.S. Srinivasan, Phase change material as a thermal energy storage material for cooling of building. J. Theor. Appl. Inf. Technol. 4 (2008)
39. S.E. Kalnæs, B.P. Jelle, Phase change materials and products for building applications: a state-of-the-art review and future research opportunities. Energy Build. **94**, 150–176 (2015)
40. T. Khadiran, M.Z. Hussein, Z. Zainal, R. Rusli, Advanced energy storage materials for building applications and their thermal performance characterization: a review. Renew. Sustain. Energy Rev. **57**, 916–928 (2016)
41. I. Marco, Seminar on phase change materials and innovation products—Brianza Plastica, Beijing, China, Oct 20 2005
42. K. Peippo, P. Kauranen, P.D. Lund, A multicomponent PCM wall optimized for passive solar heating. Energy Build. **17**, 259–270 (1991)
43. D.A. Neeper, Thermal dynamics of wallboard with latent heat storage. Sol. Energy. **68**, 393–403 (2000)
44. D. Heim, J.A. Clarke, Numerical modelling and thermal simulation of PCM–gypsum composites with ESP-r. Energy Build. **36**, 795–805 (2004)
45. D. Rozanna, A. Salmiah, T.G. Chuah, R. Medyan, S.Y. Thomas Choong, M. Sa ari, A study on thermal characteristics of phase change material (PCM) in gypsum board for building application. J. Oil Palm Res. **17**, 41 (2005)
46. H. Kaasinen, The absorption of phase change substances into commonly used building materials. Sol. Energy Mater. Sol. Cells. **27**, 173–179 (1992)
47. D. Feldman, D. Banu, D. Hawes, E. Ghanbari, Obtaining an energy storing building material by direct incorporation of an organic phase change material in gypsum wallboard. Sol. Energy Mater. **22**, 231–242 (1991)
48. Y. Zhang, G. Zhou, K. Lin, Q. Zhang, H. Di, Application of latent heat thermal energy storage in buildings: state-of-the-art and outlook. Build. Environ. **42**, 2197–2209 (2007)
49. M. Zhang, M.A. Medina, J.B. King, Development of a thermally enhanced frame wall with phase-change materials for on-peak air conditioning demand reduction and energy savings in residential buildings. Int. J. Energy Res. **29**, 795–809 (2005)
50. P. Schossig, H.-M. Henning, S. Gschwander, T. Haussmann, Micro-encapsulated phase-change materials integrated into construction materials. Sol. Energy Mater. Sol. Cells. **89**, 297–306 (2005)
51. A. Oliver, Thermal characterization of gypsum boards with PCM included: thermal energy storage in buildings through latent heat. Energy Build. **48**, 1–7 (2012)
52. E. Rodriguez-Ubinas, L. Ruiz-Valero, S. Vega, J. Neila, Applications of phase change material in highly energy-efficient houses. Energy Build. **50**, 49–62 (2012)
53. J.-F. Su, X.-Y. Wang, S.-B. Wang, Y.-H. Zhao, Z. Huang, Fabrication and properties of microencapsulated-paraffin/gypsum-matrix building materials for thermal energy storage. Energy Convers. Manag. **55**, 101–107 (2012)

54. T. Toppi, L. Mazzarella, Gypsum based composite materials with micro-encapsulated PCM: experimental correlations for thermal properties estimation on the basis of the composition. Energy Build. **57**, 227–236 (2013)
55. C. Castellon, M. Nogués, J. Roca, M. Medrano, L.F. Cabeza, *Microencapsulated Phase Change Materials (PCM) for Building Applications* (ECOSTOCK, NJ, 200)
56. N.A. Yahay, H. Ahmad, Numerical investigation of indoor air temperature with the application of PCM gypsum board as ceiling panels in buildings. Procedia Eng. **20**, 238–248 (2011)
57. S. Scalat, D. Banu, D. Hawes, J. Parish, F. Haghighata, D. Feldman, Full scale thermal testing of latent heat storage in wallboard. Sol. Energy Mater. Sol. Cells. **44**, 49–61 (1996)
58. F. Kuznik, J. Virgone, J.-J. Roux, Energetic efficiency of room wall containing PCM wallboard: a full-scale experimental investigation. Energy Build. **40**, 148–156 (2008)
59. L.F. Cabeza, C. Castellon, M. Nogues, M. Medrano, R. Leppers, O. Zubillaga, Use of microencapsulated PCM in concrete walls for energy savings. Energy Build. **39**, 113–119 (2007)
60. P. Arce, C. Castellón, A. Castell, L.F. Cabeza, Use of microencapsulated PCM in buildings and the effect of adding awnings. Energy Build. **44**, 88–93 (2012)
61. A.M. Thiele, G. Sant, L. Pilon, Diurnal thermal analysis of microencapsulated PCM-concrete composite walls. Energy Convers. Manag. **93**, 215–227 (2015)
62. H. Inaba, P. Tu, Evaluation of thermophysical characteristics on shape-stabilized paraffin as a solid-liquid phase change material. Heat Mass Transf. **32**, 307–312 (1997)
63. M. Xiao, B. Feng, K. Gong, Preparation and performance of shape stabilized phase change thermal storage materials with high thermal conductivity. Energy Convers. Manag. **43**, 103–108 (2002)
64. Y. Hong, G. Xin-Shi, Preparation of polyethylene–paraffin compound as a form-stable solid-liquid phase change material. Sol. Energy Mater. Sol. Cells. **64**, 37–44 (2000)
65. M. Xiao, B. Feng, K. Gong, Thermal performance of a high conductive shape-stabilized thermal storage material. Sol. Energy Mater. Sol. Cells. **69**, 293–296 (2001)
66. M. Li, Z. Wu, M. Chen, Preparation and properties of gypsum-based heat storage and preservation material. Energy Build. **43**, 2314–2319 (2011)
67. K. Biswas, J. Lu, P. Soroushian, S. Shrestha, Combined experimental and numerical evaluation of a prototype nano-PCM enhanced wallboard. Appl. Energy **131**, 517–529 (2014)
68. Y.P. Zhang, K.P. Lin, R. Yang, H.F. Di, Y. Jiang, Preparation, thermal performance and application of shape-stabilized PCM in energy efficient buildings. Energy Build. **38**, 1262–1269 (2006)
69. G. Zhou, Y. Zhang, K. Lin, W. Xiao, Thermal analysis of a direct-gain room with shape-stabilized PCM plates. Renew. Energy **33**, 1228–1236 (2008)
70. Y. Sun, S. Wang, F. Xiao, D. Gao, Peak load shifting control using different cold thermal energy storage facilities in commercial buildings: a review. Energy Convers. Manag. **71**, 101–114 (2013)
71. G. Kumaresan, R. Sridhar, R. Velraj, Performance studies of a solar parabolic trough collector with a thermal energy storage system. Energy **47**, 395–402 (2012)
72. C. Prieto, R. Osuna, A.I. Fernández, L.F. Cabeza, Thermal storage in a MW scale. Molten salt solar thermal pilot facility: plant description and commissioning experiences. Renew. Energy **99**, 852–866 (2016)
73. S. Guillot, A. Faik, A. Rakhmatullin, J. Lambert, E. Veron, P. Echegut, C. Bessada, N. Calvet, X. Py, Corrosion effects between molten salts and thermal storage material for concentrated solar power plants. Appl. Energy **94**, 174–181 (2012)
74. V. Zipf, A. Neuhäuser, C. Bachelier, R. Leithner, W. Platzer, Assessment of different PCM storage configurations in a 50 MWel CSP plant with screw heat exchangers in a combined sensible and latent storage–simulation results. Energy Procedia **69**, 1078–1088 (2015)
75. M.H. Mahfuz, A. Kamyar, O. Afshar, M. Sarraf, M.R. Anisur, M.A. Kibria, R. Saidur, I. Metselaar, Exergetic analysis of a solar thermal power system with PCM storage. Energy Convers. Manag. **78**, 486–492 (2014)

76. K. Bhagat, S.K. Saha, Numerical analysis of latent heat thermal energy storage using encapsulated phase change material for solar thermal power plant. Renew. Energy **95**, 323–336 (2016)
77. S.C. Lin, H.H. Al-Kayiem, Evaluation of copper nanoparticles–Paraffin wax compositions for solar thermal energy storage. Sol. Energy **132**, 267–278 (2016)
78. D.K. Singh, S. Suresh, H. Singh, B.A.J. Rose, S. Tassou, N. Anantharaman, Myo-inositol based nano-PCM for solar thermal energy storage. Appl. Therm. Eng. **110**, 564–572 (2017)
79. P.A. Galione, C.D. Pérez-Segarra, I. Rodríguez, S. Torras, J. Rigola, Multi-layered solid-PCM thermocline thermal storage for CSP. Numerical evaluation of its application in a 50 MWe plant. Sol. Energy **119**, 134–150 (2015)
80. I. IEA-ETSAP, Thermal energy storage: technology brief e17 (2013)
81. L. Miró, J. Gasia, L.F. Cabeza, Thermal energy storage (TES) for industrial waste heat (IWH) recovery: a review. Appl. Energy **179**, 284–301 (2016)
82. J. Selvaraj, M. Thenarasu, S. Aravind, P. Ashok, Waste heat recovery from castings and scrap preheating by recovered heat using an intermediate heat transfer medium, in: *Applied Mechanics and Materials* (Trans Tech Publications 2015), pp. 776–781
83. N. Maruoka, K. Sato, J. Yagi, T. Akiyama, Development of PCM for recovering high temperature waste heat and utilization for producing hydrogen by reforming reaction of methane. ISIJ Int. **42**, 215–219 (2002)
84. L. Zhang, T. Akiyama, How to recuperate industrial waste heat beyond time and space. Int. J. Exergy **6**, 214–227 (2009)
85. V. Pandiyarajan, M. Chinnappandian, V. Raghavan, R. Velraj, Second law analysis of a diesel engine waste heat recovery with a combined sensible and latent heat storage system. Energy Policy **39**, 6011–6020 (2011)
86. M. Chinnapandian, V. Pandiyarajan, A. Prabhu, R. Velraj, Experimental investigation of a cascaded latent heat storage system for diesel engine waste heat recovery, Energy Sources, Part A Recover. Util. Environ. Eff. **37**, 1308–1317 (2015)
87. T. Steinparzer, M. Haider, A. Fleischanderl, A. Hampel, G. Enickl, F. Zauner, Heat exchangers and thermal energy storage concepts for the off-gas heat of steelmaking devices. J. Phys. Conf. Ser. 12158 (2012). IOP Publishing
88. S.S. Prabu, M.A. Asokan, A study of waste heat recovery from diesel engine exhaust using phase change material. Int. J. Chem. Tech. Res. **8**, 711–717 (2015)
89. N. Gopal, R. Subbarao, V. Pandiyarajan, R. Velraj, Thermodynamic analysis of a diesel engine integrated with a PCM based energy storage system. Int. J. Thermodyn. **13**, 15–21 (2010)
90. P. Kauranen, T. Elonen, L. Wikström, J. Heikkinen, J. Laurikko, Temperature optimisation of a diesel engine using exhaust gas heat recovery and thermal energy storage (diesel engine with thermal energy storage). Appl. Therm. Eng. **30**, 631–638 (2010)
91. J. Shon, H. Kim, K. Lee, Improved heat storage rate for an automobile coolant waste heat recovery system using phase-change material in a fin–tube heat exchanger. Appl. Energy **113**, 680–689 (2014)
92. F. Baldi, C. Gabrielii, F. Melino, M. Bianchi, A preliminary study on the application of thermal storage to merchant ships. Energy Procedia. **75**, 2169–2174 (2015)
93. D.G. Beshore, F.A. Jaeger, E.M. Gartner, Thermal energy storage/waste heat recovery application in the cement industry, in: *Proceedings of First Industrial Energy Technology Conference*, pp. 747–756 (1979)
94. R. De Boer, S.F. Smeding, P.W. Bach, G. De Joode, Heat storage systems for use in an industrial batch process (Results of) a case study, in: Contribution to The Tenth International Conference on Thermal Energy Storage (ECOSTOCK, 2006)
95. M. Barati, S. Esfahani, T.A. Utigard, Energy recovery from high temperature slags. Energy **36**, 5440–5449 (2011)
96. T. Arunkumar, Y. Ao, Z. Luo, L. Zhang, J. Li, D. Denkenberger, J. Wang, Energy efficient materials for solar water distillation-A review. Renew. Sustain. Energy Rev. **115**, 109409 (2019)

97. N.A.S. Elminshawy, F.R. Siddiqui, G.I. Sultan, Development of a desalination system driven by solar energy and low grade waste heat. Energy Convers. Manag. **103**, 28–35 (2015)
98. V.G. Gude, Energy storage for desalination processes powered by renewable energy and waste heat sources. Appl. Energy. **137**, 877–898 (2015)
99. M. Faegh, M.B. Shafii, Experimental investigation of a solar still equipped with an external heat storage system using phase change materials and heat pipes. Desalination **409**, 128–135 (2017)
100. S.W. Sharshir, G. Peng, L. Wu, F.A. Essa, A.E. Kabeel, N. Yang, The effects of flake graphite nanoparticles, phase change material, and film cooling on the solar still performance. Appl. Energy **191**, 358–366 (2017)
101. T. Arunkumar, A.E. Kabeel, Effect of phase change material on concentric circular tubular solar still-Integration meets enhancement. Desalination **414**, 46–50 (2017)
102. G. Wang, Y. Fu, A. Guo, T. Mei, J. Wang, J. Li, X. Wang, Reduced graphene oxide–polyurethane nanocomposite foam as a reusable photoreceiver for efficient solar steam generation. Chem. Mater. **29**, 5629–5635 (2017)
103. G. Ni, N. Miljkovic, H. Ghasemi, X. Huang, S.V. Boriskina, C.-T. Lin, J. Wang, Y. Xu, M.M. Rahman, T. Zhang, Volumetric solar heating of nanofluids for direct vapor generation. Nano Energy **17**, 290–301 (2015)
104. O. Neumann, A.S. Urban, J. Day, S. Lal, P. Nordlander, N.J. Halas, Solar vapor generation enabled by nanoparticles. ACS Nano **7**, 42–49 (2012)
105. Y. Liu, S. Yu, R. Feng, A. Bernard, Y. Liu, Y. Zhang, H. Duan, W. Shang, P. Tao, C. Song, A bioinspired, reusable, paper-based system for high-performance large-scale evaporation. Adv. Mater. **27**, 2768–2774 (2015)
106. Y. Ito, Y. Tanabe, J. Han, T. Fujita, K. Tanigaki, M. Chen, Multifunctional porous graphene for high-efficiency steam generation by heat localization. Adv. Mater. **27**, 4302–4307 (2015)
107. H. Ghasemi, G. Ni, A.M. Marconnet, J. Loomis, S. Yerci, N. Miljkovic, G. Chen, Solar steam generation by heat localization. Nat. Commun. **5**, ncomms5449 (2014)
108. J.J. Yew, S. Rathi, S. Krueger, L.G. Bolton, J.M. Ananny, Connecting multiple accessories to a portable computing device (2011)
109. A. Van Heugten, Systems, devices, and methods for managing camera focus (2017)
110. Z. Ling, Z. Zhang, G. Shi, X. Fang, L. Wang, X. Gao, Y. Fang, T. Xu, S. Wang, X. Liu, Review on thermal management systems using phase change materials for electronic components, Li-ion batteries and photovoltaic modules. Renew. Sustain. Energy Rev. **31**, 427–438 (2014)
111. Y. Deng, C. Feng, E. Jiaqiang, H. Zhu, J. Chen, M. Wen, H. Yin, Effects of different coolants and cooling strategies on the cooling performance of the power lithium ion battery system: a review. Appl. Therm. Eng. **142**, 10–29 (2018)
112. D. Ouyang, M. Chen, Q. Huang, J. Weng, Z. Wang, J. Wang, A review on the thermal hazards of the lithium-ion battery and the corresponding countermeasures. Appl. Sci. **9**, 2483 (2019)
113. P. Huang, A. Verma, D.J. Robles, Q. Wang, P. Mukherjee, J. Sun, Probing the cooling effectiveness of phase change materials on lithium-ion battery thermal response under overcharge condition. Appl. Therm. Eng. **132**, 521–530 (2018)
114. W.G. Alshaer, S.A. Nada, M.A. Rady, E.P. Del Barrio, A. Sommier, Thermal management of electronic devices using carbon foam and PCM/nano-composite. Int. J. Therm. Sci. **89**, 79–86 (2015)
115. A. Arshad, H.M. Ali, M. Ali, S. Manzoor, Thermal performance of phase change material (PCM) based pin-finned heat sinks for electronics devices: Effect of pin thickness and PCM volume fraction. Appl. Therm. Eng. **112**, 143–155 (2017)
116. Z. Luo, H. Cho, X. Luo, K. Cho, System thermal analysis for mobile phone. Appl. Therm. Eng. **28**, 1889–1895 (2008)
117. Z.A. Qureshi, H.M. Ali, S. Khushnood, Recent advances on thermal conductivity enhancement of phase change materials for energy storage system: a review. Int. J. Heat Mass Transf. **127**, 838–856 (2018)
118. H.M. Ali, A. Saieed, W. Pao, M. Ali, Copper foam/PCMs based heat sinks: an experimental study for electronic cooling systems. Int. J. Heat Mass Transf. **127**, 381–393 (2018)

119. B. Agostini, M. Fabbri, J.E. Park, L. Wojtan, J.R. Thome, B. Michel, State of the art of high heat flux cooling technologies. Heat Transf. Eng. **28**, 258–281 (2007)
120. M. Shtein, R. Nadiv, M. Buzaglo, O. Regev, Graphene-based hybrid composites for efficient thermal management of electronic devices. ACS Appl. Mater. Interfaces **7**, 23725–23730 (2015)
121. J. Krishna, P.S. Kishore, A.B. Solomon, Heat pipe with nano enhanced-PCM for electronic cooling application. Exp. Therm. Fluid Sci. **81**, 84–92 (2017)
122. M. Itani, N. Ghaddar, K. Ghali, D. Ouahrani, W. Chakroun, Cooling vest with optimized PCM arrangement targeting torso sensitive areas that trigger comfort when cooled for improving human comfort in hot conditions. Energy Build. **139**, 417–425 (2017)
123. Y. Yusufoglu, T. Apaydin, S. Yilmaz, H.O. Paksoy, Improving performance of household refrigerators by incorporating phase change materials. Int. J. Refrig. **57**, 173–185 (2015)
124. S. Bakhshipour, M.S. Valipour, Y. Pahamli, Parametric analysis of domestic refrigerators using PCM heat exchanger. Int. J. Refrig. **83**, 1–13 (2017)
125. N. Chaiyat, Energy and economic analysis of a building air-conditioner with a phase change material (PCM). Energy Convers. Manag. **94**, 150–158 (2015)
126. M.A. Said, H. Hassan, Effect of using nanoparticles on the performance of thermal energy storage of phase change material coupled with air-conditioning unit. Energy Convers. Manag. **171**, 903–916 (2018)
127. M. Alimohammadi, Y. Aghli, E.S. Alavi, M. Sardarabadi, M. Passandideh-Fard, Experimental investigation of the effects of using nano/phase change materials (NPCM) as coolant of electronic chipsets, under free and forced convection. Appl. Therm. Eng. **111**, 271–279 (2017)
128. B. Praveen, S. Suresh, Experimental study on heat transfer performance of neopentyl glycol/CuO composite solid-solid PCM in TES based heat sink. Eng. Sci. Technol. An Int. J. **21**, 1086–1094 (2018)
129. Z. Huang, N. Xie, X. Zheng, X. Gao, X. Fang, Y. Fang, Z. Zhang, Experimental and numerical study on thermal performance of Wood's alloy/expanded graphite composite phase change material for temperature control of electronic devices. Int. J. Therm. Sci. **135**, 375–385 (2019)
130. A. Farzanehnia, M. Khatibi, M. Sardarabadi, M. Passandideh-Fard, Experimental investigation of multiwall carbon nanotube/paraffin based heat sink for electronic device thermal management. Energy Convers. Manag. **179**, 314–325 (2019)
131. B. Praveen, S. Suresh, V. Pethurajan, Heat transfer performance of graphene nano-platelets laden micro-encapsulated PCM with polymer shell for thermal energy storage based heat sink. Appl. Therm. Eng. **156**, 237–249 (2019)
132. P.M. Congedo, M. Malvoni, M. Mele, M.G. De Giorgi, Performance measurements of monocrystalline silicon PV modules in South-eastern Italy. Energy Convers. Manag. **68**, 1–10 (2013). https://doi.org/10.1016/j.enconman.2012.12.017
133. M. Chandrasekar, S. Suresh, T. Senthilkumar, M. Ganesh Karthikeyan, Passive cooling of standalone flat PV module with cotton wick structures. Energy Convers. Manag. **71**, 43–50 (2013). https://doi.org/10.1016/j.enconman.2013.03.012
134. a Q. Malik, M. Fauzi, Performance of single crystal silicon photovoltaic module in Bruneian Climate, pp. 179–188 (2010)
135. H. Jiang, L. Lu, K. Sun, Experimental investigation of the impact of airborne dust deposition on the performance of solar photovoltaic (PV) modules. Atmos. Environ. **45**, 4299–4304 (2011). https://doi.org/10.1016/j.atmosenv.2011.04.084
136. D.M. Chapin, C.S. Fuller, G.L. Pearson, A new silicon p-n junction photocell for converting solar radiation into electrical power. J. Appl. Phys. **25**, 676–677 (1954)
137. S.R. Kurtz, Opportunities and challenges for development of a mature concentrating photovoltaic power industry (2009)
138. A. Luque, S. Hegedus, *Handbook of photovoltaic science and engineering* (Wiley, 2011)
139. P. Huen, W.A. Daoud, Advances in hybrid solar photovoltaic and thermoelectric generators. Renew. Sustain. Energy Rev. **72**, 1295–1302 (2017)
140. P.G.V. Sampaio, M.O.A. González, R.M. de Vasconcelos, M.A.T. dos Santos, J.C. de Toledo, J.P.P. Pereira, Photovoltaic technologies: mapping from patent analysis. Renew. Sustain. Energy Rev. **93**, 215–224 (2018)

141. K. Agroui, A.H. Arab, M. Pellegrino, Indoor and outdoor photovoltaic modules Performances based on thin films solar cells **14**, 469–480 (2011)
142. J.Y. Ye, K. Ding, T. Reindl, A.G. Aberle, Outdoor PV module performance under fluctuating irradiance conditions in tropical climates. Energy Procedia **33**, 238–247 (2013). https://doi.org/10.1016/j.egypro.2013.05.064
143. L.S. Pantić, T.M. Pavlović, D.D. Milosavuević, A practical field study of performances of solar modules at various positions in Serbia. Therm. Sci. **19**, S511–S523 (2015). https://doi.org/10.2298/TSCI140313081P
144. M.E. Başoılu, A. Kazdaloılu, T. Erfidan, M.Z. Bilgin, B. Cąkir, Performance analyzes of different photovoltaic module technologies under¯zmit, Kocaeli climatic conditions. Renew. Sustain. Energy Rev. **52**, 357–365 (2015). https://doi.org/10.1016/j.rser.2015.07.108
145. R. Eke, H. Demircan, Performance analysis of a multi crystalline Si photovoltaic module under Mugla climatic conditions in Turkey. Energy Convers. Manag. **65**, 580–586 (2013). https://doi.org/10.1016/j.enconman.2012.09.007
146. C. Cañete, J. Carretero, M. Sidrach-de-Cardona, Energy performance of different photovoltaic module technologies under outdoor conditions. Energy **65**, 295–302 (2014). https://doi.org/10.1016/j.energy.2013.12.013
147. M.R. Abdelkader, F. Sharaf, A comparative Analysis of the Performance of Monocrystalline and Multiycrystalline PV Cells in Semi Arid Climate Conditions : the Case of Jordan. Jordan J. Mech. Ind. Eng. **4**, 543–552 (2010)
148. N. Amin, C.W. Lung, K. Sopian, A practical field study of various solar cells on their performance in Malaysia. Renew. Energy **34**, 1939–1946 (2009). https://doi.org/10.1016/j.renene.2008.12.005
149. O.M. Midtgard, T.O. Sætre, G. Yordanov, A.G. Imenes, C.L. Nge, A qualitative examination of performance and energy yield of photovoltaic modules in southern Norway. Renew. Energy **35**, 1266–1274 (2010). https://doi.org/10.1016/j.renene.2009.12.002
150. S. Dubey, J.N. Sarvaiya, B. Seshadri, Temperature dependent photovoltaic (PV) efficiency and its effect on PV production in the world—a review. Energy Procedia **33**, 311–321 (2013). https://doi.org/10.1016/j.egypro.2013.05.072
151. E. Skoplaki, J.A. Palyvos, On the temperature dependence of photovoltaic module electrical performance: a review of efficiency/power correlations. Sol. Energy **83**, 614–624 (2009). https://doi.org/10.1016/j.solener.2008.10.008
152. N. Suwapaet, P. Boonla, The investigation of produced power output during high operating temperature occurrences of monocrystalline and amorphous photovoltaic modules. Energy Procedia **52**, 459–465 (2014). https://doi.org/10.1016/j.egypro.2014.07.098
153. P. Singh, S.N. Singh, M. Lal, M. Husain, Temperature dependence of I-V characteristics and performance parameters of silicon solar cell. Sol. Energy Mater. Sol. Cells. **92**, 1611–1616 (2008). https://doi.org/10.1016/j.solmat.2008.07.010
154. S. Mekhilef, R. Saidur, M. Kamalisarvestani, Effect of dust, humidity and air velocity on efficiency of photovoltaic cells. Renew. Sustain. Energy Rev. **16**, 2920–2925 (2012). https://doi.org/10.1016/j.rser.2012.02.012
155. B. Zalba, J.M. Marın, L.F. Cabeza, H. Mehling, Review on thermal energy storage with phase change: materials, heat transfer analysis and applications. Appl. Therm. Eng. **23**, 251–283 (2003)
156. M. Mani, R. Pillai, Impact of dust on solar photovoltaic (PV) performance: Research status, challenges and recommendations. Renew. Sustain. Energy Rev. **14**, 3124–3131 (2010). https://doi.org/10.1016/j.rser.2010.07.065
157. M. Fuentes, G. Nofuentes, J. Aguilera, D.L. Talavera, M. Castro, Application and validation of algebraic methods to predict the behaviour of crystalline silicon PV modules in Mediterranean climates. Sol. Energy **81**, 1396–1408 (2007). https://doi.org/10.1016/j.solener.2006.12.008
158. M. Torres-Ramírez, G. Nofuentes, J.P. Silva, S. Silvestre, J.V. Muñoz, Study on analytical modelling approaches to the performance of thin film PV modules in sunny inland climates. Energy **73**, 731–740 (2014). https://doi.org/10.1016/j.energy.2014.06.077

159. G. Nofuentes, M. Fuentes, J. Aguilera, J.V. Muñoz, An assessment on simple modeling approaches to the electric behavior of two cis pv modules in a sunny climate. J. Sol. Energy Eng. Trans. ASME **131**, 0310131–03101310 (2009). https://doi.org/10.1115/1.3142800
160. U.K. Mirza, M. Mercedes Maroto-Valer, N. Ahmad, Status and outlook of solar energy use in Pakistan. Renew. Sustain. Energy Rev. **7**, 501–514 (2003). https://doi.org/10.1016/j.rser.2003.06.002
161. I. Ulfat, F. Javed, F.A. Abbasi, F. Kanwal, A. Usman, M. Jahangir, F. Ahmed, Estimation of solar energy potential for Islamabad, Pakistan. Energy Procedia **18**, 1496–1500 (2012). https://doi.org/10.1016/j.egypro.2012.05.166
162. D.B. Richardson, L.D.D. Harvey, Strategies for correlating solar PV array production with electricity demand. Renew. Energy **76**, 432–440 (2015). https://doi.org/10.1016/j.renene.2014.11.053
163. B. Parida, S. Iniyan, R. Goic, A review of solar photovoltaic technologies. Renew. Sustain. Energy Rev. **15**, 1625–1636 (2011). https://doi.org/10.1016/j.rser.2010.11.032
164. H.G. Teo, P.S. Lee, M.N.A. Hawlader, An active cooling system for photovoltaic modules. Appl. Energy **90**, 309–315 (2012)
165. D.J. Yang, Z.F. Yuan, P.H. Lee, H.M. Yin, Simulation and experimental validation of heat transfer in a novel hybrid solar panel. Int. J. Heat Mass Transf. **55**, 1076–1082 (2012). https://doi.org/10.1016/j.ijheatmasstransfer.2011.10.003
166. A. Makki, S. Omer, H. Sabir, Advancements in hybrid photovoltaic systems for enhanced solar cells performance. Renew. Sustain. Energy Rev. **41**, 658–684 (2015). https://doi.org/10.1016/j.rser.2014.08.069
167. C. Schwingshackl, M. Petitta, J.E. Wagner, G. Belluardo, D. Moser, M. Castelli, M. Zebisch, A. Tetzlaff, Wind effect on PV module temperature: analysis of different techniques for an accurate estimation. Energy Procedia **40**, 77–86 (2013). https://doi.org/10.1016/j.egypro.2013.08.010
168. E. Kermani, Manifold micro-channel cooling of photovoltaic cells for high efficiency solar energy conversion systems (2008)
169. M.M. Rahman, M. Hasanuzzaman, N.A. Rahim, Effects of various parameters on PV-module power and efficiency. Energy Convers. Manag. **103**, 348–358 (2015)
170. O. Dupré, R. Vaillon, M.A. Green, Physics of the temperature coefficients of solar cells. Sol. Energy Mater. Sol. Cells. **140**, 92–100 (2015)
171. J.W. Stultz, L.C. Wen, Thermal performance testing and analysis of photovoltaic modules in natural sunlight. LSA Task Rep. **5101**, 31 (1977)
172. A. Hasan, S.J. McCormack, M.J. Huang, B. Norton, Evaluation of phase change materials for thermal regulation enhancement of building integrated photovoltaics. Sol. Energy **84**, 1601–1612 (2010)
173. C.J. Smith, P.M. Forster, R. Crook, Global analysis of photovoltaic energy output enhanced by phase change material cooling. Appl. Energy **126**, 21–28 (2014)
174. U. Stritih, Increasing the efficiency of PV panel with the use of PCM. Renew. Energy **97**, 671–679 (2016)
175. S. Maiti, S. Banerjee, K. Vyas, P. Patel, P.K. Ghosh, Self regulation of photovoltaic module temperature in V-trough using a metal–wax composite phase change matrix. Sol. Energy **85**, 1805–1816 (2011)
176. P.H. Biwole, P. Eclache, F. Kuznik, Phase-change materials to improve solar panel's performance. Energy Build. **62**, 59–67 (2013)
177. M.J. Huang, P.C. Eames, B. Norton, N.J. Hewitt, Natural convection in an internally finned phase change material heat sink for the thermal management of photovoltaics. Sol. Energy Mater. Sol. Cells. **95**, 1598–1603 (2011)
178. M.C. Browne, K. Lawlor, A. Kelly, B. Norton, S.J. Mc Cormack, Indoor characterisation of a photovoltaic/thermal phase change material system. Energy Procedia. **70**, 163–171 (2015)
179. M.C. Browne, B. Norton, S.J. McCormack, Phase change materials for photovoltaic thermal management. Renew. Sustain. Energy Rev. **47**, 762–782 (2015)

180. P. Atkin, M.M. Farid, Improving the efficiency of photovoltaic cells using PCM infused graphite and aluminium fins. Sol. Energy. **114**, 217–228 (2015)
181. A.H.A. Al-Waeli, K. Sopian, M.T. Chaichan, H.A. Kazem, A. Ibrahim, S. Mat, M.H. Ruslan, Evaluation of the nanofluid and nano-PCM based photovoltaic thermal (PVT) system: an experimental study. Energy Convers. Manag. **151**, 693–708 (2017)
182. M.K.A. Ali, H. Xianjun, L. Mai, C. Bicheng, R.F. Turkson, C. Qingping, Reducing frictional power losses and improving the scuffing resistance in automotive engines using hybrid nanomaterials as nano-lubricant additives. Wear **364**, 270–281 (2016)
183. A.H.A. Al-Waeli, H.A. Kazem, M.T. Chaichan, K. Sopian, Experimental investigation of using nano-PCM/nanofluid on a photovoltaic thermal system (PVT): technical and economic study. Therm. Sci. Eng. Prog. **11**, 213–230 (2019)
184. N.S. Dhaidan, J.M. Khodadadi, T.A. Al-Hattab, S.M. Al-Mashat, Experimental and numerical investigation of melting of NePCM inside an annular container under a constant heat flux including the effect of eccentricity. Int. J. Heat Mass Transf. **67**, 455–468 (2013)
185. D. Groulx, Numerical study of nano-enhanced PCMs: are they worth it? in *Proceedings of the 1st Thermal and Fluid Engineering Summer Conference (TFESC)*, New York, Aug 2015, pp. 9–12
186. R. Parameshwaran, K. Deepak, R. Saravanan, S. Kalaiselvam, Preparation, thermal and rheological properties of hybrid nanocomposite phase change material for thermal energy storage. Appl. Energy **115**, 320–330 (2014)
187. M. Amin, N. Putra, E.A. Kosasih, E. Prawiro, R.A. Luanto, T.M.I. Mahlia, Thermal properties of beeswax/graphene phase change material as energy storage for building applications. Appl. Therm. Eng. **112**, 273–280 (2017)
188. Z. Luo, H. Zhang, X. Gao, T. Xu, Y. Fang, Z. Zhang, Fabrication and characterization of form-stable capric-palmitic-stearic acid ternary eutectic mixture/nano-SiO2 composite phase change material. Energy Build. **147**, 41–46 (2017)
189. S. Sharma, L. Micheli, W. Chang, A.A. Tahir, K.S. Reddy, T.K. Mallick, Nano-enhanced phase change material for thermal management of BICPV. Appl. Energy. **208**, 719–733 (2017)
190. Z. Ma, W. Lin, M.I. Sohel, Nano-enhanced phase change materials for improved building performance. Renew. Sustain. Energy Rev. **58**, 1256–1268 (2016)
191. S.C. Lin, H.H. Al-Kayiem, Evaluation of copper nanoparticles—Paraffin wax compositions for solar thermal energy storage. Sol. Energy **132**, 267–278 (2016). https://doi.org/10.1016/j.solener.2016.03.004
192. N. Abdollahi, M. Rahimi, Potential of water natural circulation coupled with nano-enhanced PCM for PV module cooling. Renew. Energy **147**, 302–309 (2020)
193. H.E. Abdelrahman, M.H. Wahba, H.A. Refaey, M. Moawad, N.S. Berbish, Performance enhancement of photovoltaic cells by changing configuration and using PCM (RT35HC) with nanoparticles Al_2O_3. Sol. Energy **177**, 665–671 (2019)
194. M. Al-Jethelah, S.H. Tasnim, S. Mahmud, A. Dutta, Nano-PCM filled energy storage system for solar-thermal applications. Renew. Energy **126**, 137–155 (2018)
195. S.Y. Wu, H. Wang, S. Xiao, D.S. Zhu, An investigation of melting/freezing characteristics of nanoparticle-enhanced phase change materials. J. Therm. Anal. Calorim. **110**, 1127–1131 (2011)
196. T.-P. Teng, C.-C. Yu, Characteristics of phase-change materials containing oxide nano-additives for thermal storage. Nanoscale Res. Lett. **7**, 611 (2012)
197. T. Oya, T. Nomura, M. Tsubota, N. Okinaka, T. Akiyama, Thermal conductivity enhancement of erythritol as PCM by using graphite and nickel particles. Appl. Therm. Eng. **61**, 825–828 (2013)
198. Q. He, S. Wang, M. Tong, Y. Liu, Experimental study on thermophysical properties of nanofluids as phase-change material (PCM) in low temperature cool storage. Energy Convers. Manag. **64**, 199–205 (2012)
199. Y. Zeng, L.-W. Fan, Y.-Q. Xiao, Z.-T. Yu, K.-F. Cen, An experimental investigation of melting of nanoparticle-enhanced phase change materials (NePCMs) in a bottom-heated vertical cylindrical cavity. Int. J. Heat Mass Transf. **66**, 111–117 (2013)

200. Y. Cui, C. Liu, S. Hu, X. Yu, The experimental exploration of carbon nanofiber and carbon nanotube additives on thermal behavior of phase change materials. Sol. Energy Mater. Sol. Cells. **95**, 1208–1212 (2011)
201. M. Li, A nano-graphite/paraffin phase change material with high thermal conductivity. Appl. Energy **106**, 25–30 (2013)
202. A.H. Ali, S.I. Ibrahim, Q.A. Jawad, R.S. Jawad, M.T. Chaichan, Effect of nanomaterial addition on the thermos physical properties of Iraqi paraffin wax. Case Stud. Therm. Eng. 100537 (2019)
203. M. Kazemi, A. Kianifar, H. Niazmand, Nanoparticle loading effect on the performance of the paraffin thermal energy storage material for building applications. J. Therm. Anal. Calorim. 1–7 (n.d.)

